"I am someone who has chosen to have my family members and patients cared for by Dr. David Perlmutter. I deeply respect his open-mindedness and expertise as a physician. He combines the skills of a well-trained traditional medical practitioner and caring physician with the wisdom of one who goes beyond the limitations of today's so-called medical education and the closed-minded medicine I see practiced today. David is a leader in his field."
—Bernie Siegel, M.D.

"Dr. Perlmutter gives readers at least a ten-year head start in accessing this revolutionary information for their own health and the health of those for whom this information can be vitally important."
—Jeffrey Bland, Ph.D., president,
Institute for Functional Medicine

"A major contribution. Dr. Perlmutter has combined science and clinical experience in creating a practical resource for doctors and patients."
—Leo Galland, M.D., FACP, FACN,
author of *The Four Pillars of Healing*

"I have looked to Dr. Perlmutter for his medical wisdom many times in the past. The new volume shows how far research has come and how effectively he communicates its findings."
—Neal D. Barnard, M.D., president, Physicians
Committee for Responsible Medicine,
and author of *Eat Right, Live Longer*

"Dr. Perlmutter provides sound advice, supported by the latest and most well respected medical research."
—Russell B. Roth, M.D., past president,
American Medical Association

"Any patient will benefit from Dr. Perlmutter's advice, and all physicians should learn of these treatments if they want to do the best for their patients."
—Michael Janson, M.D., president, American
College for Advancement in Medicine, and
author of *Dr. Janson's New Vitamin Revolution*

The Better

RIVERHEAD BOOKS

New York

Brain Book

The Best Tools for Improving
Memory and Sharpness and for
Preventing Aging of the Brain

DAVID PERLMUTTER, M.D., FACN, and CAROL COLMAN

THE BERKLEY PUBLISHING GROUP
Published by the Penguin Group
Penguin Group (USA) Inc.
375 Hudson Street, New York, New York 10014, USA
Penguin Group (Canada), 10 Alcorn Avenue, Toronto, Ontario M4V 3B2, Canada
(a division of Pearson Penguin Canada Inc.)
Penguin Books Ltd., 80 Strand, London WC2R 0RL, England
Penguin Group Ireland, 25 St. Stephen's Green, Dublin 2, Ireland (a division of Penguin Books Ltd.)
Penguin Group (Australia), 250 Camberwell Road, Camberwell, Victoria 3124, Australia
(a division of Pearson Australia Group Pty. Ltd.)
Penguin Books India Pvt. Ltd., 11 Community Centre, Panchsheel Park, New Delhi—110 017, India
Penguin Books (NZ), cnr. Airborne and Rosedale Roads, Albany, Auckland 1310, New Zealand
(a division of Pearson New Zealand Ltd.)
Penguin Books (South Africa) (Pty.) Ltd., 24 Sturdee Avenue, Rosebank, Johannesburg 2196,
South Africa

Penguin Books Ltd., Registered Offices: 80 Strand, London WC2R 0RL, England

Every effort has been made to ensure that the information contained in this book is complete and accurate. However, neither the publisher nor the authors are engaged in rendering professional advice or services to the individual reader. The ideas, procedures, and suggestions contained in this book are not intended as a substitute for consulting with your physician. All matters regarding your health require medical supervision. Neither the authors nor the publisher shall be liable or responsible for any loss or damage allegedly arising from any information or suggestion in this book.

The recipes contained in this book are to be followed exactly as written. The publisher is not responsible for your specific health or allergy needs that may require medical supervision. The publisher is not responsible for any adverse reactions to the recipes contained in this book.

While the authors have made every effort to provide accurate telephone numbers and Internet addresses at the time of publication, neither the publisher nor the authors assume any responsibility for errors, or for changes that occur after publication. The publisher does not have any control over and does not assume any responsibility for author or third-party websites or their content.

The opinions expressed herein represent the personal views of the authors and not of the publisher.

First Riverhead hardcover edition: August 2004
First Riverhead trade paperback edition: August 2005
Riverhead trade paperback ISBN: 1-59448-093-1

The Library of Congress has catalogued the Riverhead hardcover edition as follows:

Perlmutter, David, M.D.
   The better brain book: the best tools for improving memory and sharpness
and for preventing aging of the brain / David Perlmutter and Carol Colman.
      p.   cm.
   Includes bibliographical references and index.
   ISBN 1-57322-278-X
   1. Memory disorders—Prevention.   2. Dietary supplements.   3. Nootropic agents.
4. Brain.   I. Colman, Carol.   II. Title.
RC394.M46P476   2004   2003069367
616.8'4—dc22

PRINTED IN THE UNITED STATES OF AMERICA

10   9   8   7   6   5   4   3   2   1

*This book is dedicated to expanding the knowledge
that devastating neurological diseases can be prevented.*

"The time to fix the roof is when the sun is shining."
—JOHN F. KENNEDY

Acknowledgments

I would like to thank Amy Hertz for her energy, intelligence, and input. She was with us every step of the way on this project, and this book would not have made it without her. We owe a debt of gratitude to Susan Petersen Kennedy for her support, and Marilyn Ducksworth, Erinn Hartman, Steve Oppenheimer, and the terrific publicity team for their creativity and hard work. Much thanks to Marc Haeringer for all of his help. He was always there when we needed him. A special thanks to Janis Vallely, my agent, whose hard work and enthusiasm helped to make this project a reality.

Thank you to Gabrielle Rabner, M.S., R.D., for designing the meal plan and offering such great advice.

I would also like to thank Bill Kelley with whom I've learned that even adults can have pals. And finally, I would also like to thank my wife, Leize, for always being there, heart and soul.

Contents

PART

1

Is Your Brain at Risk?

Am I Losing It?

- YOU'VE JUST BEEN introduced to someone at the office or at a party, and seconds later, as you start to introduce her to your friend . . . you realize you can't remember her name.
- You're standing in front of a dozen colleagues who have gathered to hear your presentation. It's your moment to shine . . . but in midsentence you stop dead. To your immense embarrassment, you find yourself asking, "What was I saying?"
- You walk into the kitchen to find your reading glasses . . . but as you enter the room, you momentarily go blank and wonder, "Now, why am I here?"
- You can't remember where you put your reading glasses . . . again.
- You pick up the phone . . . and for the third time today, you intend to dial one number when you're actually dialing another.
- You find that you're having trouble calculating restaurant tips . . . something you used to do easily in your head.

Does this sound familiar? Are you having difficulty concentrating at work, remembering names and appointments, and keeping track of your glasses and keys? Are you easily distracted and less able to stay on task? Do you nod out during long meetings? Do you feel that you're not as sharp or on-the-ball as you once were? Do you get cranky more often than you once did? Do you worry that you're so out of it that you're going to lose your next promotion to some kid right out of school? Do you worry, "If I'm like this now, what am I going to be like in ten years?" Are you tired of making jokes about "senior moments"?

In moments of desperation, do you wonder, "Am I losing it?"

The conventional wisdom is that becoming forgetful, moody, and even a little spacey is a "normal" sign of aging that happens to everyone and is no cause for concern. Moreover, you might think, even if your brain *is* turning to mush, there's nothing much you can do about it anyway, except sit back and watch with a combination of amusement and horror as you become mentally slower and duller with each passing day.

But that conventional wisdom is outdated, outmoded, and outrageously wrong. The slowdown in brain function that begins during midlife—the memory problems, difficulty focusing, brain fog, irritability, and loss of mental agility and physical coordination—is not a "normal" part of aging. It is the beginning of a downward spiral that is destroying your brain, slowly and insidiously chipping away at your ability to function at your peak and stay on the top of your game.

And believe me, it *is* something to worry about. The process of deterioration that is causing the where-did-I-put-my-keys? and what-was-I-just-doing? syndromes is the same process that can lead to serious medical conditions such as Alzheimer's disease, Parkinson's disease, and stroke. What's worse, most of us unwittingly lead lifestyles that accelerate this destructive process. We're promoting the demise of our brains. Consider this:

- If you regularly use any over-the-counter or prescription medication, you could be putting your brain at risk. Dozens of commonly used drugs, including antacids, birth control pills, nonaspirin pain relievers, and cholesterol-lowering medications can starve your brain of important nutrients that keep it functioning at an optimal level and protect against brain aging.

- If you don't routinely read food labels to screen out bad "brain-busting" ingredients, chances are you are consuming foods that make your brain sluggish and slow down your reaction time.
- If you drink unfiltered tap water, spray insecticides in your home or garden, or use underarm deodorant or dandruff shampoo, you may be exposing your brain to toxins that will age it faster.
- If you are "stressed out" on the job or at home, your brain is being constantly bathed in stress hormones that can destroy its memory center and "dumb you down."

That's the bad news. The good news is that brain degeneration is not inevitable. You can stop it, reverse it, and recover what you have lost. Your brain has amazing powers of regeneration, and you can tap into them by supplying the right raw materials. You don't have to abandon your life dreams or lower your expectations because you think that your brain can't go the distance. My Better Brain program will give you the tools you need to reinvigorate, protect, and maintain your brain. I will introduce you to my scientifically proven program of natural therapies, including a Better Brain Workout that boosts memory and mental speed, dietary strategies that can rejuvenate sluggish brain cells, and nutritional supplements that provide the fuel you need to keep your brain running on all cylinders.

I am a board-certified neurologist, a specialist in the brain and nervous system, and I have been practicing medicine for more than two decades. At my offices, the Perlmutter Health Center in Naples, Florida, I have treated thousands of patients who have come to me from around the world with problems ranging from mild midlife memory loss and brain fog to severe neurological deficits resulting from disease. Those of you who may be worried about developing neurological problems— whether because of family history, lifestyle, or just the sense that your brain is slowing down—will learn how to prevent them and preserve your brain function. Those of you who have already been diagnosed with neurological problems will learn how to arrest the progression of your disease, reverse deterioration, and restore lost function.

The Better Brain program is organized in three tiers based on your level of risk:

- *Tier 1, Prevention and Maintenance,* is a supplement regimen for those who have few risk factors but want to maintain optimum brain function throughout their entire lives. If this is you, you haven't yet reached the "where-did-I-put-my-keys?" stage, and if you follow this program, you may avoid it completely. Although you may get away with being less vigilant about your diet, at least for now, you should try to follow the Better Brain meal plan most of the time. You can also benefit from doing the Better Brain Workout at least two to three times a week; it will make your good brain even better.

- *Tier 2, Prevention, Repair, and Enhancement,* is a supplement program for those who are at *moderate* risk who would like to improve their brain performance and prevent further deterioration. If this is you, within a few weeks you will feel as sharp and alert as you did back in the days when words and ideas flowed effortlessly, and you had mental energy to spare. If you want to stay out of Tier 3, I strongly urge you to follow my meal plan and do the Better Brain Workout everyday.

- *Tier 3, Recovery and Enhancement,* is an intensive supplement regimen for those who are at *high* risk, including those who may already be experiencing a noticeable decline in mental ability. If this is you, Tier 3 is an aggressive program to help bring you back to a higher level of functioning and prevent further decline. It is imperative that you follow my meal plan and do the Better Brain Workout every day.

Which Program Is Right for You?

Should you follow Tier 1, Tier 2, or Tier 3? The Brain Audit, the self-assessment test in chapter 2, will show you which program is best suited to your needs. The Brain Audit contains 45 questions about your mental function, lifestyle, diet, environment, medications, stress level, and medical history. The Brain Audit will take only a few minutes of your time but will reveal volumes about the state of your brain health. Most people are completely unaware of the common, everyday things in their lives that can sap their brain power, accelerate brain aging, and leave them vulnerable to a neurological disease. The Brain Audit will help identify the risk factors in your life that are threatening the health of your brain so you can begin to make positive changes. Remember to consult your physician before starting any diet, supplement, or exercise program.

How to Use This Book

The Better Brain Book is divided into three parts. Part I, "Is Your Brain at Risk?" reveals the hidden risk factors that are hurting your mental performance right now and increasing your susceptibility to serious memory issues and neurological disease down the road. The Brain Audit in Chapter 2 will help you identify which tier of the Better Brain program is right for you.

Part II, "Tools for a Better Brain," includes a Brain Workout to reinvigorate your brain cells, a lifestyle guide, a meal plan, and supplement regimens for Tiers 1, 2, and 3. Chapter 11 describes four medical tests for your brain that I believe should be included in your annual physical.

Part III, "Specific Brain Disorders and What to Do About Them," is geared for those who have already been diagnosed with neurological disorders and offers treatment protocols that have been helpful for many patients. It covers Parkinson's disease, stroke, multiple sclerosis, Alzheimer's disease, vascular dementia and Lou Gehrig's disease (ALS).

THE BRAIN AUDIT that follows will help you assess the status of your brain so you can determine which tier of the Better Brain program is right for you.

The Brain Audit will reveal things about your lifestyle, home or work environment, or daily routine that could be putting the health of your brain at risk. This part of the Better Brain program helps you make constructive changes in your life that will have a profound impact on your future. Making these simple changes may not only save your brain but could very well save your life.

Please answer "Yes" or "No" to the questions that follow. When you have finished the Brain Audit, add up your score as instructed. Your score will tell you which tier of the Better Brain program is right for you.

Rating Your Mental Performance

1. Do you find that you have to write things down or you forget to do them?　　　　　　　　　　　Yes ☐　No ☐
2. Do you frequently misplace things (like your keys, wallet, or glasses)?　　　　　　　　　　Yes ☐　No ☐

3. Do you find it increasingly difficult to stay focused during a long meeting?   Yes ☐   No ☐

4. Do you worry that you won't remember someone's name shortly after being introduced?   Yes ☐   No ☐

5. Do you find that it is difficult for you to resume a task once you have been interrupted? (For example, if you stop to answer the telephone while you are doing something, do you find it difficult to remember where you left off?)   Yes ☐   No ☐

6. If the radio or television is on in a room, do you have difficulty reading or concentrating on your work?   Yes ☐   No ☐

7. Do you find it more difficult to do simple mathematical calculations in your head—like figuring out the tip on a restaurant check or keeping score during a tennis or card game?   Yes ☐   No ☐

8. Do you get more frustrated than you used to when confronted with a mental challenge?   Yes ☐   No ☐

9. Do you feel overly concerned or fearful when you are forced to learn something new, like a new computer system at work?   Yes ☐   No ☐

10. Do you find it is becoming more difficult to follow the plot in a novel or a movie?   Yes ☐   No ☐

11. Is it easier for you to remember something that happened 30 years ago than three days ago?   Yes ☐   No ☐

Your Age

12. Are you over 40 years old?   Yes ☐   No ☐
13. Are you over 50 years old?   Yes ☐   No ☐
14. Are you over 60 years old?   Yes ☐   No ☐
15. Are you over 70 years old?   Yes ☐   No ☐
16. Are you over 80 years old?   Yes ☐   No ☐

Your Diet

17. Do you eat foods with trans-fatty acids three or more times a week? (*Trans-fatty acids are found in most brands of margarine and in many processed baked*

goods and snack foods. *If you don't know the answer
to this question, you probably are!*)   Yes ☐   No ☐

18. Do you use artificially sweetened foods or beverages?   Yes ☐   No ☐

19. Is there usually more meat on your plate than
vegetables?   Yes ☐   No ☐

20. Do you drink more than two glasses of wine or two
alcoholic drinks on most days?   Yes ☐   No ☐

21. Do you eat sweets or dessert every day?   Yes ☐   No ☐

What's in Your Medicine Cabinet?

22. Do you routinely take any drugs—prescription or over the counter—that
could be harmful to your brain? *Routinely* is defined as four times a week
or more. Before you say no, keep in mind that this includes many antacids,
painkillers, heart medications, asthma treatments, and even birth control
pills. For a list of potentially brain-damaging drugs, see the list of drugs at
the end of the Brain Audit. Each drug counts as a separate Yes answer.
Write down the number of drugs that you take in the "Yes" column.
Yes ☐   No ☐

Personal Habits

23. Do you routinely get less than eight hours sleep a
night?   Yes ☐   No ☐

24. Do you use a cell phone without an earphone?   Yes ☐   No ☐

25. Do you have a sedentary lifestyle? (That is, you don't
exercise regularly, defined as 30 minutes of
moderately vigorous exercise, such as brisk walking,
running, or weight training at least three times a
week.)   Yes ☐   No ☐

26. Do you smoke tobacco or have you smoked tobacco
within the last 20 years?   Yes ☐   No ☐

27. Have you ever used cocaine?   Yes ☐   No ☐

Your Environment

28. Do you drink well water?   Yes ☐   No ☐

29. Have you ever lived in a place where the house or
yard was routinely treated for insects?   Yes ☐   No ☐

30. Do you live in a house or an apartment building constructed before 1978?   Yes ☐   No ☐
31. Do you sleep with an electric blanket or clock radio within 3 feet of your head?   Yes ☐   No ☐

Your Stress Level

32. Are you going through a stressful time in your life— for example, is your job stressful or are you in the midst of a divorce, out of work, having financial concerns, or caring for a sick relative?   Yes ☐   No ☐
33. Are you all work and no play—that is, you rarely engage in leisure activities?   Yes ☐   No ☐
34. Do you come from a single-parent home?   Yes ☐   No ☐
35. Do you have 3 or more older siblings?   Yes ☐   No ☐
36. Did you lose a parent during childhood or teenage years?   Yes ☐   No ☐
37. Did you experience physical or emotional abuse as a teen or child?   Yes ☐   No ☐
38. Did you serve in the military during wartime?   Yes ☐   No ☐

Your Medical History

39. Do you have a parent, grandparent, or sibling who has suffered from a neurological disease such as Alzheimer's, Parkinson's, or senile dementia or who has had a stroke?   Yes ☐   No ☐
40. Are you more than 20 pounds overweight?   Yes ☐   No ☐
41. Have you been diagnosed with either Type 1 or Type 2 diabetes?   Yes ☐   No ☐
42. Have you ever been diagnosed with depression?   Yes ☐   No ☐
43. Have you ever experienced a head trauma that resulted in a loss of consciousness?   Yes ☐   No ☐
44. Do you have a history of coronary artery disease?   Yes ☐   No ☐
45. Do you have treated or untreated high blood pressure with either a systolic reading (the top number) of over 150 or a diastolic reading (the bottom number) of over 85?   Yes ☐   No ☐

Are You Taking Any of These Drugs?

Antacids and Stomach Acid Suppressors

GENERIC NAME: aluminum hydroxide (with magnesium carbonate, or magnesium hydroxide, magnesium hydroxide and simethicone, or magnesium trisilicate)

DRUG TYPE: antacid. BRAND NAMES: Gaviscon, Aludrox, Di-Gel, Gelusil, Maalox, Magalox, Mylanta

GENERIC NAME: Cimetidine. BRAND NAME: Tagamet

GENERIC NAME: Famotidine. BRAND NAME: Pepcid

GENERIC NAME: Lansoprazole. BRAND NAME: Prevacid

GENERIC NAME: Nizatidine. BRAND NAME: Axid

GENERIC NAME: Omeprazole. BRAND NAME: Prilosec

GENERIC NAME: Ranitidine. BRAND NAMES: Zantac, Zantac 75

Pain Relievers

GENERIC NAME: aspirin (acetylsalicylic acid). BRAND NAMES: Pure aspirin is sold under numerous brand names, including many private store labels. Percodan and Empirin are combination aspirin and codeine products.

Nonaspirin Pain Relievers

GENERIC NAME: Acetaminophen. BRAND NAMES: Panadol, Tylenol, Tylenol Arthritis Pain, Aspirin-Free Anacin, plus countless house brands. Acetaminophen may also be added to over-the-counter cold medicines and codeine-containing pain relievers.

Antidepressants

GENERIC NAME: Amitriptyline. BRAND NAME: Elavil

GENERIC NAME: Desipramine. BRAND NAME: Norpramin

GENERIC NAME: Doxepin. BRAND NAME: Sinequan

GENERIC NAME: Imipramine. BRAND NAME: Tofranil

GENERIC NAME: Nortriptyline. BRAND NAMES: Aventil, Pamelor

GENERIC NAME: Protriptyline. BRAND NAME: Vivactil

Antipsychotic Drugs

GENERIC NAME: Haloperidol. BRAND NAME: Haldol

Blood Pressure–Lowering Drugs

GENERIC NAME: Atenolol. BRAND NAME: Tenormin

GENERIC NAME: Bisoprolol. BRAND NAME: Zebeta

GENERIC NAME: Bumetanide. BRAND NAME: Bumex

GENERIC NAME: Clonidine. BRAND NAME: Catapres

GENERIC NAME: Furosemide. BRAND NAME: Lasix

GENERIC NAME: Hydralazine. BRAND NAME: Apresoline

GENERIC NAME: Hydroclorothiazide (HCTZ): diuretic used alone or in combination with other medications to lower blood pressure. BRAND NAMES: Aldactazide, Capozide, Combipres, Dyazide, HydroDIURIL, Hyzaar, Lopressor-HCT, Lotensin HCT, Maxide, Microzide, Moduretic, Prinzide, Vaseretic, Zestoretic

GENERIC NAME: Metoprolol. BRAND NAMES: Lopressor, Toprol

GENERIC NAME: Nadolol. BRAND NAME: Corgard

GENERIC NAME: Pindolol. BRAND NAME: Visken

GENERIC NAME: Propranolol. BRAND NAME: Inderal

GENERIC NAME: Torsemide. BRAND NAME: Demadex

GENERIC NAME: Triamterene. BRAND NAMES: Dyrenium, or in combination with other drugs in Maxide and Dyazide

Cholesterol-Lowering Drugs

GENERIC NAME: Atorvastatin. BRAND NAME: Lipitor

GENERIC NAME: Cholestyramine. BRAND NAME: Colestid

GENERIC NAME: Fluvastatin. BRAND NAME: Lescol

GENERIC NAME: Lovastatin. BRAND NAME: Mevacor

GENERIC NAME: Pravastatin. BRAND NAME: Pravachol

GENERIC NAME: Simvastatin. BRAND NAME: Zocor

Antidiabetic Drugs

GENERIC NAME: Glipizide. BRAND NAME: Glucotrol

GENERIC NAME: Glyburide. BRAND NAMES: DiaBeta, Glynase, Micronase

GENERIC NAME: Metformin. BRAND NAME: Glucophage

GENERIC NAME: Tolazemide. BRAND NAME: Tolinase

Asthma Drugs

GENERIC NAME: Beclomethasone (oral inhaler). BRAND NAME: Vanceril

GENERIC NAME: Budesonide (oral inhaler). BRAND NAME: Pulmacort

GENERIC NAME: Budesonide (nasal inhaler). BRAND NAME: Rhinocort

GENERIC NAME: Flunisolide (nasal inhaler). BRAND NAME: Nasalide

GENERIC NAME: Flunisolide (oral inhaler). BRAND NAME: Aerobid

GENERIC NAME: Fluticasone (oral inhaler). BRAND NAME: Flovent

GENERIC NAME: Mometasone (nasal inhaler). BRAND NAME: Nasonex

GENERIC NAME: Theophylline. BRAND NAME: Aerolate

GENERIC NAME: Triamcinolone (oral inhaler). BRAND NAME: Azmacort

Antibiotics

GENERIC NAME: Trimethoprim; frequently prescribed for chronic urinary tract infections. BRAND NAMES: Bactrim, Septra

Anticonvulsant Drugs; Also Prescribed as Antidepressants

GENERIC NAME: Carbamazepine. BRAND NAME: Tegretol

GENERIC NAME: Ethosuximide. BRAND NAME: Zarontin

GENERIC NAME: Fosphenytoin. BRAND NAME: Cerebyx

GENERIC NAME: Mephobarbital. BRAND NAME: Mebaral

GENERIC NAME: Phenobarbital. BRAND NAME: Phenobarbital

GENERIC NAME: Phenytoin. BRAND NAME: Dilantin

GENERIC NAME: Primidone. BRAND NAME: Mysoline

GENERIC NAME: Valproic Acid. BRAND NAMES: Depakote, Depakene

Anti-Parkinson's Drugs

GENERIC NAME: Carbidopa and Levodopa (combination of two drugs). BRAND NAME: Sinemet

Corticosteroids: Antiinflammatory Drugs

These are used to treat asthma, arthritis, allergy, and pain.

GENERIC NAME: Methyl Prednisolone. BRAND NAME: Medrol

GENERIC NAME: Prednisone. BRAND NAMES: Deltasone, Orasone

Estrogens

There are dozens of different brands of estrogen products, and I have only listed the major brands here. If you are taking estrogen in the form of birth control pills or hormone replacement therapy, even if you don't see your particular estrogen drug on the list, assume that it works just like other estrogen and can cause nutrient depletion. (See chapter 4.)

GENERIC NAME: Estrogens (with or without progesterone) as contraceptives, oral or skin patch. BRAND NAMES: Ortho-Novum, OrthoTri-Cyclen, Ovral, Ovcon, Demulen, Loestrin (any estrogen-containing oral contraceptive)

GENERIC NAME: Estrogens (with or without progesterone) for hormone replacement therapy for menopause or hysterectomy, oral, skin patch, or cream. BRAND NAMES: Premarin, Prempro, Activela, Combinpatch, Estrotest, any hormone replacement product containing estrogen.

Estrogen Substitutes for Osteoporosis

GENERIC NAME: Raloxifene. BRAND NAME: Evista

Nonsteroidal Antiinflammatory Drugs (NSAIDs)

GENERIC NAME: Celecoxib. BRAND NAME: Celebrex

GENERIC NAME: Ibuprofen. BRAND NAMES: Advil, Bayer Select, Motrin, Midol, etc.

GENERIC NAME: Indomethacin. BRAND NAME: Indocin

GENERIC NAME: Naproxen. BRAND NAMES: Naprosyn, Aleve, etc.

Assessing Your Level of Risk

Each "Yes" answer is worth 1 point. Add up all your "Yes" answers to get your final tally. For question 22, each of the drugs that you take routinely counts as one "Yes" answer, so you accrue several points for this question.

Which Program Should You Follow?

Total Score 0–6

Tier 1 | If you scored between 0 and 6 on the Brain Audit, you should follow the Tier 1 program for *Prevention and Maintenance*. Tier 1 is for people who are basically healthy, have reasonably good health habits, and do not have any specific problems. My hunch is that if you have scored in Tier 1, you are 30 years old or younger, and you are likely to breathe a sigh of relief and say, "Well, I'll just leave well enough alone, and not do anything." That could be the mistake of your life. The pathological process that is the root cause of brain degeneration is just beginning. Over the next decade, the destruction will only accelerate, and if you do nothing to stop it, you will soon find yourself in Tier 2 or Tier 3. The Tier 1 program is fairly simple; I recommend that you take a few key nutrition supplements (see chapter 6) and that you follow the Better Brain meal plan. Although the Brain Workout in chapter 9 is optional for you, you can still benefit from it. It will sharpen your memory and increase your reaction time; in other words, it will make your good brain even better.

Total Score 7–30

Tier 2 | If you scored between 7 and 30 on the Brain Audit, you should follow Tier 2, for *Prevention, Repair, and Enhancement.* Tier 2 is for people who are at moderate risk and would like to improve their brain performance and prevent further problems. Just as your muscles need to be exercised so they don't get flabby, so do your brain cells. The Brain Workout (see chapter 9) will help restore mental agility and improve your ability to process information and retrieve it on demand. The Better Brain meal plan will keep your brain well nourished with the raw materials required to build a better brain within weeks. The Tier 2 supplement regimen in chapter 6 will replenish key nutrients in your brain that decline with age and are your brain's best protection against further damage. Be vigilant about implementing the lifestyle changes suggested in chapter 7, and review the list of drugs in your Brain Audit. Paying close attention to your lifestyle and medications can help prevent your minor problems from turning into serious ones.

Total Score Above 30

Tier 3 | If you scored over 30 on the Brain Audit, you should follow Tier 3, *Recovery and Enhancement.* Tier 3 is for people who are at *high* risk and/or may already be experiencing a noticeable decline in mental ability. You need to take all aspects of the Better Brain program very seriously. I strongly urge you to be extra careful about following the dietary recommendations and taking your Tier 3 supplements daily. On one positive note, you should quickly see an improvement in cognitive function, especially if you do your Brain Workout every day. I also suggest that you review chapter 11, "Four Medical Tests That Can Save Your Brain."

What do your risk factors really mean? Are some more serious than others? To learn the answers to these questions, turn to chapter 3.

Understanding Your Risk Factors

Your Mental Performance: Questions 1–11

If you answered yes to any of the questions in "Rating Your Mental Performance," you're probably wondering, "Why am I having such trouble remembering simple things such as names and numbers? Why am I constantly misplacing things? Why am I having difficulty learning new tasks? Does this mean that I'm on the road to early senility? What does my mental performance have to do with the rest of the questions on the Brain Audit?"

First, let me assure you that even if you answered Yes for every question on the mental function list, you are not senile or ready to be turned out to pasture. If you were, you would never be able to read this book or answer the questionnaire! But that doesn't mean that your brain function is what it should be, or, more important, what it *can* be; and it is negatively affecting the quality of your life. Even if you answered yes to just one or two of these questions, you are not operating at your mental peak. You are experiencing a loss in brain power. Here's what your answers on this part of the Brain Audit can tell you about your brain.

Memory Loss

It is most likely that you answered yes to the questions related to memory loss, because short-term memory is typically the first to go (questions 1, 2, 4, and 11). You forget names immediately after being introduced to someone, you put down the car keys and can't find them a few minutes later, and if you don't write something down, you won't remember to do it. You may be surprised to learn that your real problem is not one of memory—your brain knows where you've put the keys and the time of your doctor's appointment—your problem is that you can't retrieve this information on demand. Eventually, you do remember where you put the keys, often after a few anguished minutes of searching for them, and a day or two later you may even recall that you skipped your doctor's appointment. But its frustrating not to be able to have this information at your fingertips when you need it.

Memory loss typically gets progressively worse over time. If you are having difficulty remembering recent events but can clearly remember things that happened decades earlier (see question 11), you have a more advanced form of memory loss. This doesn't happen in a vacuum; it's caused by a combination of age, poor diet, nutrient deprivation, and physical illness, and can often be reversed with proper intervention. Those who answer yes to question 11 are likely to score in the high Tier 2 or even Tier 3 range.

Staying on Task

Did you answer yes to question 5, "Do you find that it is difficult for you to resume a task once you have been interrupted?" Did you answer yes to question 6, "If the radio or television is on in a room, do you have difficulty reading or concentrating on your work?" It's not your imagination—you are not as focused as you used to be, and background "noise" that you were once able to ignore can now hamper your mental performance. Why? It's harder for you to do more than one thing at a time. To paraphrase an oft-cited computer term, you are having trouble *multitasking*. Here's what I mean. I have two teenage kids, and I am often amazed to see how they can listen to the radio, talk on the phone, chat with their friends on their computer, and do their homework without missing a beat. They can do all of these activities simultaneously because the young brain is capable of multitasking and can absorb lots of information very quickly. An older brain gets "stuck" along the way. For example, we've all had the experience of walking into a room and forgetting why we went in there. We remember the first step of the task, walking into the room, but we forget the second part of the task, which is why we went into the room in the first place. Do you have trouble taking notes in a meeting and following the speaker? This is another example of a failure in multitasking, where your brain is "overloaded" with information and can't process it quickly enough.

Difficulty Learning

Once the "brain drain" has begun, learning new materials can also be more of a challenge. Did you answer yes to questions 8, 9, and 10, which all relate to absorbing and learning new material? Your brain is losing some of its "hard drive," which makes it more difficult to assimilate new information. For example, it may be harder to learn a new language or more difficult to keep track of a complicated plot in a novel (which also requires some multitasking skills and mental focus). Here's what's happening. Your brain is made up of billions of cells called neurons, and every neuron is encased in a protective cell membrane. Each neuron has tiny, branchlike connections called dendrites. Dendrites are very important for learning. When your brain is stimulated by a challenging task, you make new dendrites, which enhances communication between brain cells and keeps you smarter. Although you can make new dendrites well into old age, it gets more difficult as you get older, which is why it's tougher to keep things in your head, like a new language, or the plot of a novel with lots of twists and turns.

Brain Fatigue

Is your brain running out of steam? Difficulty doing simple mathematical calculations in your head or keeping score at a tennis game (question 7) is a sign of tired brain. You are losing mental stamina, and you may have difficulty with tasks requiring extended periods of concentration (like writing a report at work.) A yes response to question 7 is also a sign that your reaction time is slowing down (often due to brain fatigue). You don't process information as quickly as you used to, and chances are you don't think as quickly on your feet as you used to either.

Your Brain Is Under Attack

Why is your brain on a downward spiral? Why can't your brain keep functioning like it did at its peak when you could soak up new information like a sponge, recall facts in nanoseconds, and stay alert for hours on end? Why are you spending half the day hunting down your glasses or the car keys? There are forces at work within your brain that, if left to their own devices, will ultimately destroy it. And everything that you do in your life, from your sleep habits, to the medication that you take, to what you put on your plate, can either halt this process or fast-forward it.

What is this insidious process that is targeting your brain?

The same forces that are aging your body are aging your brain, only they hit your brain earlier and harder. These culprits are at the core of virtually all brain problems, from mild memory issues to brain fog to severe Alzheimer's disease. They are: (1) the proliferation in the brain of destructive chemicals called free radicals, and (2) the decline in the ability of brain cells to make energy. As I will show, these factors are closely related to each other, and their effects on brain function are profound.

Your brain is a hotbed of activity—literally. It is the most metabolically active organ of the body. It uses 20 percent of the oxygen you consume to make the energy to fuel all of its activities. Energy is made in the specialized parts of the cell called the mitochondria. There is a price to pay for making all of this energy. Every time a cell makes energy—any cell, in any part of the body—it also produces toxic substances called free radicals. Think of it as cellular pollution. Free radicals are unstable; they don't keep to themselves, and they like to bond with other molecules in healthy cells. When they do, they release energy, or "heat," that can damage surrounding tissues and organs, such as the heart, joints, and skin. This process is called oxidation. Given the choice, free radicals would rather bond with fat cells than with any other kind of cell. This is a problem because our brain and nerve cells are made mostly out of fat— particularly the cell membrane, the protective covering of the cell. The cell membrane is the most important part of the brain cell because it is where most of the brain's work is done. Every time you learn something new, think, create, or speak, it involves brain cell membranes. Over time, unless arrested, free radicals can destroy substantial amounts of brain and nerve tissue through this process of oxidation.

If free radicals are allowed to run amok, they can cause another major headache for the brain: they interfere with its ability to make energy. They target particular fat-rich parts of the cell membrane, the mitochondria, the energy-producing centers of the cell. Think of them as the cell's power plants. When the mitochondria of your brain cells are injured, they become less efficient, produce less energy, and increase free radical production. You fatigue more easily, you can't concentrate as well as you used to, and you are more vulnerable to the damaging effects of stress.

A brain that is sharp and on the ball is a brain in which brain cells can communicate easily with each other. When you ask "Where did I put the keys?" you want your brain to respond instantaneously. Once again, free radicals clog up this vital process. Your brain cells talk to each other by releasing chemicals called neurotransmitters. They are the oil that keeps your brain running smoothly, and some are involved in particular functions. For example, one neurotransmitter called acetylcholine is involved in memory and learning. Another neurotransmitter, dopamine, is involved in balance and physical movement. Yet another neurotransmitter, serotonin, regulates mood and appetite. The right levels of neurotransmitters are critical for a well-functioning brain. Free radicals can inhibit the brain's ability to produce neurotransmitters, which will have a profound impact on memory, learning, mood, and even balance and hand-eye coordination.

Free radicals pose another potentially deadly problem for the brain—they promote inflammation. High levels of free radicals spark a defensive response by the immune system, which sends out cells to attack what it sees as "invaders." This creates even more free radicals, and more injured cells, less energy for your brain, and more inflammation! Like a fire out of control, inflammation can spread throughout your body. In recent years, inflammation has been linked to nearly all chronic brain diseases, including Parkinson's disease, Alzheimer's disease, multiple sclerosis, and dementia.

So what do free radicals have to do with misplacing your keys? Everything! The cells in the hippocampus, the memory center of the brain, are especially vulnerable to free radical attack. That's why memory loss or, more precisely, the inability to retrieve information on demand is one of the first signs of brain aging. What do free radicals have to do with your ability to learn new tasks, or grasp new ideas or concepts? Everything! Damaged brain cells do not learn or store information as efficiently as healthy brain cells. The demands on one's brain don't decline, but the capacity of one's brain to cope with the demands does decline. Mental activities that once came easily, such as remembering names and dates, finding the right word, or processing new information, become progressively harder.

Rescue Your Brain

How can you stop this free radical assault on your brain so that you can regain your brain power? The human body has developed an elaborate system for keeping free radicals in check and preventing the damage that results from oxidation. This system is known as the "antioxidant defense system" because it produces chemical compounds, known as "antioxidants," that function as the free radical police of the body. They capture and arrest free radicals before they can cause damage through oxidation. There are hundreds of naturally occurring antioxidants. Some are produced by the body, while others must be obtained from food—

particularly fruits and vegetables—or from supplements. One antioxidant in particular, glutathione, which is produced by the body, is especially important for the health of the brain and nervous system. Coenzyme Q10 is another antioxidant produced by the body that protects brain cells from free radicals and inflammation. Some of the most important brain-protective antioxidants, such as vitamin E, are not produced by the body and must be obtained through supplements. The problems is that as we age, our antioxidant defense system loses some of its punch. We stop producing enough glutathione, Co-Q10, and other antioxidants to adequately protect the brain. It's no coincidence that the decline of our antioxidant defense system precisely coincides with the onset of mild cognitive deficits, such as memory loss and difficulty with concentrating or learning new tasks.

Your lifestyle can aggravate the situation. Chronic stress, whether emotional or physical, stimulates the production of free radicals. Being overweight also promotes free radical production, and so does lack of sleep. So do common chemicals present in food additives, pesticides, and environmental pollutants. Some of these substances not only enhance the production of free radicals and inflammation but destroy key antioxidants that protect the brain, such as glutathione and Co-Q10. Some of the worst offenders are over-the-counter and prescription drugs that we take routinely, like acetaminophen (Tylenol), cholesterol-lowering drugs, antacids, and blood pressure–lowering drugs. To compound the problem, most people do not eat a diet that is rich in brain-protective antioxidants or take antioxidant supplements to fill the gap.

Are you doing enough to protect your brain against the ravages of free radical attack and inflammation?

The rest of the Brain Audit reveals things about your lifestyle, home or work environment, or daily routine that may be promoting free radical production and putting your brain at risk. Undoubtedly you would like to know more about your results. Here I explain what these risk factors really mean and how they can impact your life. You *can save your brain,* but first you need to understand what's wrong so that you can begin to make it right.

Your Age And Your Brain: Questions 12–16

By age 40, about two-thirds of all people experience some mental decline, which can accelerate with each decade of life. The longer you've lived, the more likely it is that your brain will be damaged by free radicals and inflammation, and the more likely you are to see a decrease in mental function. The downward spiral begins with the typical mild memory problems or brain fog so common in midlife and can accelerate exponentially with each decade. By age 65, 1 out of 100 people will have symptoms of dementia, such as confusion, severe forgetfulness, and an inability to manage on their own. By age 75 this is true for 1 out of 10, and by age 85, 1 out of 2. If you don't stop the spiral, you can go from not remembering where you *put* your keys to not knowing how to *use* your keys.

Clearly, age plays a role in brain and neurological function, but these statistics don't tell the whole story. When it comes to impaired brain function, age alone is not the primary culprit. Simply growing older doesn't mean that you will grow frail in body and mind. If you are free of other risk factors, it is possible to maintain maximum brain function *at any age*. Your health and lifestyle are far more important in terms of assessing your risk of neurological problems than your age. On the Brain Audit, a healthy octogenarian can score in the low-risk range if he or she does not have any other risk factors other than advanced age, while a much younger person with a risky lifestyle and/or medical problems can score in the high-risk range.

Your Diet: Questions 17–21

What you put on your plate every day can have a huge impact on the health of your brain tomorrow. If you answered yes to any of the questions in this section, you are eating and drinking things that are harming your brain. Cleaning up your diet is a major part of the Better Brain program. The right diet can be a powerful tool to promote healing and enhance brain function. The wrong diet can promote brain degeneration and accelerate brain aging.

Trans Fats and Saturated Fats

This may be the most important question on the Brain Audit. If you are not aware of the importance of dietary fat and brain function, chances are you are eating the wrong fat and destroying your brain cells. Your brain is made up primarily of fat. Where does all that fat come from? Quite simply, the fat that ends up in your brain is from the fat you consume in your diet. Some types of fat (especially fat found in fatty fish) are great for your brain. These fats can enhance overall brain function and are especially important for maintaining a good mood. Other types of fat, however, are terrible for the brain. The worst fats of all are transfatty acids, which are commonly found in many brands of margarine, processed baked goods, fried foods, and saturated fats, which are found in animal products (meat and full-fat dairy products like butter). These fats not only promote inflammation but prevent good fats from getting into your brain cells. Both trans-fatty acids and saturated fats can make your brain cells hard and rigid and interfere with the brain's ability to process information quickly. As I tell my patients, if you eat a diet high in "sluggish fat," you will have a "sluggish brain." For more information on how to get bad fat out of your body, and good fat into your body, see chapter 5.

Artificial Sweeteners

Aspartame, a commonly used artificial sweetener (found in diet soda and other diet foods) may be toxic to your brain. Aspartame contains chemicals called excitotoxins that can cross the brain/blood barrier and overstimulate brain cells, which disrupts the normal production of neurotransmitters and promotes free radicals. In susceptible people, excitotoxins may trigger headache and mood swings and may even promote the growth of brain tumors. There are two other commonly used food additives that also contain excitotoxins: MSG and hydrolyzed vegetable protein. For more information on excitotoxins, see page 153.

More Meat Than Vegetables

If you're not filling up your plate with vegetables, you are missing out on important antioxidants found in plants (also in fruit) that can shield your brain from the damaging effects of free radicals. Please turn to chapter 5 to see which fruits and vegetables offer the best brain protection.

Alcohol—Two Your Health!

Alcohol is the most widely used drug in the world, and its health effects, like those of caffeine, are dose related. One or two servings of any kind of alcoholic beverage daily (one serving equals 3 ounces of wine, one standard shot glass of spirits, or one 12-ounce bottle of beer) reduces the risk of neurological disease, but more than two glasses of alcohol daily increases the risk. Alcohol contains beneficial antioxidants that protect against heart disease, stroke, and even some forms of cancer. The problem is, alcohol also contains toxic chemicals that can cause free radicals. Alcohol is first purified by the liver, the body's prime detoxifying organ, before it is absorbed by the body. More than two glasses of alcohol can place a heavy burden on the liver and deplete glutathione, the brain's primary antioxidant. Alcohol should not be used in any amount by people who have liver problems or who are on those medications that should not be mixed with alcohol (such as acetaminophen and many of the NSAIDs).

The Sugar Connection

Can't remember where you put the candy bar? If you're having trouble remembering things and are generally forgetful, it could be due to sugar overload. People who eat sweets regularly (which I define as two or more servings daily of sugar-laden desserts like a cup of ice cream, two cookies, a slice of cake, a glass of soda, or even a small bag of chips) are putting the health of their brains at risk. Excess sugar consumption increases the risk of elevated levels of blood sugar, which in turn increases the risk of developing memory problems at an earlier age than normal. Eating sweets could also increase your risk of neurological diseases. A recent study of patients with Parkinson's disease found that they ate larger quantities of sweet foods and more snacks than their healthy peers. (For a list of sweets, snacks, and sweeteners you should avoid, see chapter 5.)

What's In Your Medicine Cabinet: Question 22

Unbeknownst to most laypeople and even many physicians, numerous commonly used over-the-counter and prescription drugs can decrease brain function and make you more susceptible to neurological problems. Even the best of drugs can have nasty side effects, including robbing the body of vital nutrients that protect against disease. Many drugs deplete the brain of key antioxidants such as Co-Q10 and glutathione. Other drugs sap the body of B vitamins, which help control homocysteine, an amino acid (building block of protein) that, if allowed to rise to unhealthy levels, can promote inflammation and blockage of blood vessels. High levels of homocysteine are a risk factor for depression, memory problems, and a slowdown in cognitive function. Other commonly used drugs—including many over-the-counter antacids—contain aluminum, a heavy metal that also promotes inflammation and may increase the risk of Alzheimer's disease. The very medicines that you think are protecting your health are promoting free radical production and inflammation. (For more information, see chapter 4.)

A word about drugs in general. We live in a culture where every ache and pain, physical or emotional, is treated with a pill. I caution against the overuse of drugs of any kind, prescription, over the counter, or recreational. A drug that is powerful enough to cause a physiological change in your body has the potential to cause negative side effects. In particular, many drugs deplete the body of important nutrients that help protect against free radicals and inflammation. I'm not saying not to use pharmaceutical drugs ever; there are times that taking the right medicine can be lifesaving, and I prescribe medication to patients when appropriate. I am cautioning, however, against the casual use of drugs, particularly when there may be better options, such as making positive lifestyle changes like improving your diet, getting enough rest, and embarking on an exercise program.

Personal Habits: Questions 23–27

The things that you do every day, or even on occasion, can have a dramatic impact on the health of your brain. If you checked yes to any of these questions, your personal habits may be putting your brain at risk.

Missing Sleep

According to a recent study by the National Sleep Foundation, as many as 47 million Americans are routinely sleep deprived, that is, they regularly do not get the optimal 7–8 hours sleep a night, and often get by on much less. Missing even one night of sleep can have an immediate effect on your physical and cognitive function. Sleep-deprived people are typically tired and confused and do not score as well on mental acuity tests. They also suffer from important biochemical changes that can be detrimental to health, including a surge in stress hormones that can damage brain cells in the hippocampus, the memory center of the brain. (See chapter 7 for information on how to get a good night's sleep.)

Cell Phones

There are 170 million cell phone users in the United States. I have one myself, but I use it rarely. I'm not anti–cell phones, but I am worried about the long-term health effects on the brain of constant cell phone use. Every time you use a cell phone, you are sending radio waves into your brain. Depending on how close the cell phone antenna is to the head, your brain could be absorbing a significant amount of electromagnetic energy. No one knows whether this will prove to be harmful or not, but there is evidence that constant bombardment of the brain with electromagnetic radiation may damage brain cell DNA and increase the risk of brain cancer and brain degeneration. Newer models of cell phones with earphones that minimize the contact with your head are safer. Nevertheless, the consequences of long-term use of cell phones, particularly for people who use them as their primary phone, are still unknown. I use mine only for emergencies, and I am particularly alarmed at the high use of cell phones among children and teenagers. For more information on how to use cell phones safely, see chapter 8.

Lack of Exercise

Ever notice how after spending the day hunched over a computer or sitting on the couch watching TV, you feel tired and dull? The only way to shake off the lethargic feeling is to take a brisk walk or work out at the gym. Even though you're tired, once you begin to move, the brain fog lifts, and you become more focused. Physical activity increases blood flow to the brain, which gives your brain cells more nourishment and makes you feel instantly alert. If you routinely don't get enough exercise, not only will you experience a decline in brain function in the short run but your brain will suffer in the long run. Canadian researchers followed more than 4,600 men and women over the age of 65 who did not have Alzheimer's disease to see how many would eventually develop the disease. Five years later, 194 people in the study had been diagnosed with Alzheimer's. The researchers compared the lifestyle and habits of those who developed Alzheimer's to those who did not. Regular exercise proved to be the most important protective factor, even more important than a family history of the disease. In fact, regular physical activity lowered the risk of Alzheimer's by as much as 30 percent!

Smoking Tobacco

There are 40 million smokers in the United States, and each one of them is at greater risk of dementia, stroke, heart disease, diabetes, and some forms of cancer. All of these can have a detrimental effect on brain function.

What makes tobacco smoke so deadly? Each puff of a cigarette produces thousands of free radicals that can overwhelm the body's antioxidant defenses. Smoking damages your lungs by promoting inflammation, which eats away at delicate lung tissue over time. What you may not realize is that inflammation from your lungs can spread to your brain and can damage brain cells. Furthermore, smokers have notoriously low levels of brain-protective antioxidants such as glutathione and vitamin E, which will leave your brain cells even more vulnerable to free radical attack and inflammation. To add insult to injury, smoking destroys blood vessels, which can interfere with the flow of blood to your brain, robbing it of essential nutrients. Don't smoke! Even if you quit smoking, you still have to undo the damage inflicted on your body, and you must be vigilant about minimizing your exposure to other free radical–generating toxins. You must take your antioxidant supplements and be very careful about living a clean, "toxin-free" life. (See chapter 6 for more information.)

Cocaine and Your Brain

The cocaine high is a result of a surge in dopamine production in the brain. An ongoing study being conducted at the University of Michigan suggests that cocaine may destroy the dopamine-producing cells in the brain, which could increase the risk of Parkinson's disease later in life. Cocaine users are also at high risk of hemorrhagic stroke, which is characterized by bleeding within the brain. This is a no-brainer—don't use this drug. If you have used cocaine at any time in your life, be vigilant about reducing your risk of stroke (maintain normal blood pressure and don't put on excess weight).

Your Environment: Questions 28–31

Many people who are conscientious about living a healthy lifestyle are shocked to learn that they are being exposed to hidden toxins every day, at work or at home, that can cause neurological damage. In chapter 8, I give specific information on how you can eliminate as many toxins from your environment as possible. Some of the worst offenders are as follows.

Well Water

You may think that the well water in your country house is purer than the city water from a municipal water supply, and you'd be wrong. Well water is often tainted with pesticide runoff from local farms. Pesticides are powerful nerve toxins; farmers and other people routinely exposed to pesticides have a dramatically higher risk of Parkinson's disease. Pesticides promote the formation of free radicals in the body and challenge the body's antioxidant defenses.

Pesticides in Your Home

If you've lived in a home that was (or still is) routinely treated with pesticides indoors or in the yard, you may have been exposed to nerve toxins and could be at greater risk of developing Parkinson's disease. Some people are more susceptible to the deleterious effects of pesticides than others, but they should be avoided by everyone whenever possible. I also recommend that people eat organic produce—grown without pesticides—to reduce pesticide exposure. Taking antioxidant supplements can also protect your brain cells against the free radicals produced by these toxins.

Houses Built Before 1978

Many private homes and apartment buildings constructed prior to 1978 contain vestiges of lead paint (which can be inhaled) or lead in plumbing pipes that can leach into drinking water. Lead is a known neurotoxin that is particularly dangerous to children, but I have also seen my share of adults affected by lead exposure. When otherwise healthy people come to me with symptoms such as confusion or signs of nerve damage, I order a simple blood test to determine whether or not they have been exposed to lead. If they have, I recommend a procedure called chelation therapy, an intravenous nutrient therapy that removes the lead from their body. I also put them on the Tier 3 supplement regimen to chase the free radicals out of their body. In 1978, Congress outlawed the use of lead in construction, but there are lots of old homes that still contain lead. For more information on how to test your home for lead and how to protect yourself against the toxic effects of lead, see chapter 8.

Electric Blankets, Clock Radios: The EMF Threat

Electricity emits invisible forces called electromagnetic fields (EMFs). (So do cell phones, as I mentioned earlier.) Electromagnetic fields radiate from electric blankets, computers, TV screens, clock radios, and anywhere electricity flows. There is a high concentration of EMFs in electrical delivery systems such as power lines, transformers, or high-tension electrical wires. Some, but not all, studies have shown a link between living or working in close proximity to these electrical delivery systems and some forms of cancer, including brain tumors. Although EMFs can disrupt the production of antioxidants in the body and promote free radicals, there is some controversy as to whether or not they are harmful. Given the preponderance of electrical devices in our lives today, I think it is wise to protect yourself against excess EMF exposure, and I talk about this more in chapter 8. Don't worry, I'm not going to suggest that you give up the conveniences of modern life; but there are ways that you can use them safely.

Your Stress Level: Questions 32–38

People tend to dismiss stress as a minor annoyance or a purely "emotional" problem. In reality, chronic stress can have a profoundly negative effect on your body as well as your mind. Chronic stress alters the chemistry of your brain and the body, revving up the production of free radicals. Over time, stress can raise your blood pressure and elevate blood sugar levels, which increases the risk of diabetes, heart attack, and stroke.

Many people may not even realize that their lives are stressful or that they live and work in a particularly stressful environment.

Stress in Your Life

Even short-term exposure to stress hormones can result in temporary memory problems and poorer scores on mental function tests. At work, stress could be interfering with job productivity and performance, not to mention the negative health implications. Stress in your personal life is no less lethal. People who experience trauma or a very stressful event early in their lives may be at much greater risk of developing Alzheimer's disease down the road. Given the complexity of life in the twenty-first century, it's unrealistic to suggest that you can rid your life of all stress, but it is possible to learn how to manage it better so that it doesn't become toxic.

All Work and No Play

Having a good time is good for your brain. First, recreational activities are good stress relievers, which will help reduce the toxic effect of stress on your brain and your body. Chances are that when you're engaged in an activity that you truly enjoy, you are thinking about something other than your day-to-day problems. Second, some recreational activities, especially those that require different skills from those you use every day, may encourage the growth of new neurons, which help keep your brain young and active. Third, engaging in recreational activities can help maintain friendships, which is critical for mental well-being. People who regularly partake in leisure activities are also at reduced risk of Alzheimer's disease and dementia, probably for all of the reasons just mentioned. The bottom line is, if all you do is work, you are cheating your brain out of important stimulation.

Single-Parent Homes

If you were raised in a single-parent home, you are at an increased risk of developing Alzheimer's disease and other forms of dementia. The reason is fairly obvious: single-parent homes are typically under greater economic and social stress than homes headed by two parents.

Born After Three Siblings

Your birth order can affect the state of your brain later in life. If you were born after three siblings, you are at greater risk for developing Alzheimer's disease. Why? Researchers speculate that younger kids who have to constantly compete for parental attention with older siblings may find that to be very stressful.

Loss of a Parent

Losing a parent during your childhood or teenage years puts you at increased risk of Alzheimer's disease, for obvious reasons. The untimely death of a parent is a traumatic event and is stressful for both the children and surviving parent and can have long-lasting effects.

History of Abuse Early in Life

Emotional or physical abuse during childhood or teenage years is a risk factor of Alzheimer's disease. Once again, the culprit is stress. Abuse creates an intolerably stressful environment that can be lethal to your brain cells, not just while it's happening but long after the fact. If you have a history of either emotional or physical abuse, it is critical that you get the right psychological counseling to help you better cope with this trauma. Maintaining good health habits, stress reduction, and keeping your antioxidant defenses strong can help prevent, repair, and preserve your brain function.

Active Military Service

Serving in the military during a war can be especially stressful and can increase the risk of health problems, including neurological disease. Soldiers are not only subjected to high levels of physical and emotional stress but are more likely to be exposed to toxic chemicals. Anyone who has served in the armed forces needs to be especially careful about adopting a healthy lifestyle and avoiding further toxic exposure.

Your Medical History: Questions 39–45

Family History

If you have a family history of neurological disease, you'll be relieved to know that genetics is only a small part of the story. Just the fact that your parent or grandparent may have had a neurological disease doesn't mean that you are destined to have the same problem. In most cases, the genetic link is relatively weak; nevertheless, it should not be ignored.

A genetic risk for a specific problem is often due to another problem that creates an environment within the body that promotes a particular disease. You don't necessarily inherit a specific gene for a particular neurological disease, but you may inherit a seemingly unrelated problem that can leave you vulnerable to a neurological condition. The so-called Alzheimer's gene—the ApoE4 gene—is a prime example of this phenomenon. ApoE is a protein found in the body that is involved in the transport of cholesterol. There are three different types of the ApoE protein: ApoE2, ApoE3, and ApoE4. ApoE2 and ApoE3 are considered good because they function as important antioxidants in the brain. In contrast, the ApoE4 variety does not offer any meaningful antioxidant protection, which makes it less desirable than the other two.

You can inherit any combination of ApoE genes from your parents. Under ideal conditions, you would have two ApoE 2 or 3 genes. Not everyone wins at the genetic roll of the dice. Some people don't inherit these protective genes but instead carry one or two ApoE4 genes, which offer no antioxidant protection. As a result they are at greater risk of developing Alzheimer's disease and other neurological problems. Even if you have two ApoE4 genes, however, it does not mean that you are automatically programmed to develop Alzheimer's disease no matter what you do. Carrying an ApoE4 gene is a sign that you may have a weaker-than-normal antioxidant defense system, which in turn may leave your brain defenseless against free radical attack and lead to Alzheimer's disease. Fortunately, despite your genetic situation, you can take supplemental antioxidants to strengthen your antioxidant defenses to "save your brain." (Of course, you won't know whether or not you have this gene if you are not tested. Please read chapter 11 on why I think this test is so important.)

The bottom line is that genetics plays a role in neurological disease, but I do not consider it to be a major player. If you know that you have a family history of neurological problems, it's important for you to be vigilant about maintaining and monitoring your brain health. It is critical for you to follow a healthy lifestyle and to be particularly careful about avoiding medical problems that could increase your risk of neurological disease.

Obesity

If you are significantly overweight or obese (weigh 20 percent more than your ideal body weight) during midlife, you are at greater risk of having memory problems and experiencing signs of brain aging earlier than normal. If you are a man who is obese, you are at even greater risk of losing your brain power. According to the Framingham Heart Study, which tracked the health and lifestyles of thousands of people after 1950, obese men scored as much as 23 percent lower on mental function tests than nonobese men. For reasons that are not fully understood, obese women did not score lower on mental function tests. Whether you are male or female, being obese also increases your risk of developing Parkinson's and Alzheimer's disease in later life, and having a stroke at any age.

Why does obesity have such a negative effect on the brain?

- Obese people are more likely to have elevated levels of blood sugar, which has been linked to memory loss.
- Obesity makes every organ in your body work harder, which stresses the body, and promotes inflammation and free radical production. Free radicals and inflammation can damage your delicate brain cells

and leave you with fewer brain cells with which to think, learn, and work.

- Obesity increases the risk of diabetes, heart disease, and high blood pressure, all associated with a substantial increased risk of brain disorders.

Obesity is primarily caused by poor eating habits—too many high-calorie, nutrient-deficient foods and too few antioxidant-rich foods. There is a cure: change what you are putting on your plate. (For more information on the right way to eat, see chapter 5.)

Diabetes

Diabetics and people with prediabetic conditions (such as high blood sugar) have poorer scores on memory tests than healthy people of the same age. Why? High levels of glucose react with specific proteins in the brain, forming substances called advanced glycation end products (AGES), which can destroy proteins throughout the body and in the brain and create more free radicals.

Depression

People with depression may have lower levels of key neurotransmitters in the brain that are essential for the maintenance of mood, memory, and learning. I'm not suggesting that you run to your doctor for a prescription for an antidepressant! Diet, nutrition, and stress control techniques can help relieve depression and restore normal brain chemistry. One good fat in particular, DHA, found in food and available in supplement form, is essential for good mood and a healthy brain and has been shown to relieve depression. (See page 111 for more information.) Getting enough sleep and reducing stress is also helpful, and I refer you to chapter 7 for tips on how to change your lifestyle to improve your brain and your mood. People with a history of depression are also at high risk of Alzheimer's disease. Why? Depression is linked to high levels of homocysteine, an amino acid that promotes inflammation in blood vessels and is associated with a significantly higher risk of dementia and heart disease. Having elevated homocysteine levels is also an independent risk factor for Alzheimer's. (I recommend that everyone have their homocysteine levels checked each year as part of their annual physical. See chapter 11.)

Head Trauma

Head trauma that results in a loss of consciousness can increase your risk of Alzheimer's disease, Parkinson's, and dementia. The culprit is chronic inflammation. A head injury can produce immediate problems such as deficits in short-term memory, having difficulty finding the right word (and often substituting the wrong word), persistent headaches, and slow reaction time. Even when these obvious problems subside, the inflammation in the brain persists and, over time, can increase the risk of neurological disease. High-intensity athletes are at particular risk of having a traumatic head injury that can cause problems for them right after they occur and later in life. Many accidents that lead to head trauma are entirely preventable simply by wearing the right protective gear. This is especially important for young athletes. (See page 139 for information on how to protect yourself against the most common causes of head trauma.)

Heart Disease

Coronary artery disease (CAD) is a condition characterized by the buildup of plaque in arteries delivering blood to the heart, which can restrict flow throughout the body and to your brain. This increases your risk of developing neurological problems, notably memory loss and depression. Heart disease is also linked to elevated homocysteine, high levels of inflammation, and too few antioxidants, the same risk factors that can accelerate brain degeneration.

High Blood Pressure

If you are showing signs of forgetfulness or are dismissing lapses in memory as "senior moments," you should have your blood pressure checked. High pressure can interfere with blood flow to the brain, and according to a recent study reported at the American Heart Association's Fifty-Seventh Annual Blood Pressure Research Conference, may be the underlying cause of memory problems for many people. High blood pressure is defined as a systolic reading (the top number) of over 140 or a diastolic reading (the bottom number) of over 85. About 20 percent of the U.S. population has high blood pressure, and one-third of them don't even know it. If you want to keep your brain sharp and alert, make sure that you have your blood pressure checked by your physician each year and, if you have high blood pressure, that it is medically treated. High blood pressure also increases your risk of heart disease and stroke and is often associated with elevated homocysteine levels. As we know by now, excess homocysteine wildly accelerates inflammation and destroys brain cells. High blood pressure can also cause atrophy or shrinkage of the brain, which can have a profound impact on brain function, ranging from depression to concentration problems to loss of memory.

Maintaining good health and taking the appropriate steps to bolster your antioxidant defenses and reduce inflammation will help preserve your brain function. All of the problems listed in this chapter can be vastly improved, if not cured, by proper nutrition, the right lifestyle, and nutritional supplements.

Are you taking any drugs that could put the health of your brain at risk? Turn to chapter 4 for more information on drugs that deplete your brain of vital, "brain-saving" nutrients and what you can do to protect yourself.

How Common Prescription and Over-the-Counter Drugs Can Be Hazardous to Your Brain

ARE YOU TAKING birth control pills? Do you pop antacids every time you feel a twinge in your tummy? Are you taking an antidepressant? Are you using a drug to lower your cholesterol? Do you take acetaminophen every day for arthritis? If you're using one or more prescription or over-the-counter drugs, you could be interfering with your brain's ability to function at optimal levels and putting the health of your brain at risk. If you're getting forgetful, have low mental energy, are moody and irritable, or feel that you're not as sharp as you used to be, your symptoms could be caused or aggravated by any number of prescription or over-the-counter drugs.

Commonly used drugs—from stomach acid suppressors to antidepressants, to birth control pills, to cholesterol-lowering medicines, to pain relievers—can deplete your brain of life-saving nutrients that protect against free radical damage and can contain ingredients that can promote inflammation. Drug-induced nutrient depletion can hurt any organ system, but I am primarily concerned about drugs that can harm the brain. There are drugs that lower levels of Co-Q10 or glutathione, two critical antioxidants for the brain that protect against free radical damage and inflammation. Moreover, dozens of drugs destroy B vitamins, which are essential because they help control the amino acid homocysteine, which in excess can promote inflammation and kill brain cells.

In this chapter I list some commonly used prescription and over-the-counter drugs that can sap the body of key brain nutrients. Read over the lists carefully to see if you are taking any of these medications. Please note that, given the hundreds of drugs on the market and the new ones brought to the market every day, I could not include every possible drug. As you review the list, however, you will see that entire categories of drugs may deplete the body of a specific nutrient. For example, estrogen interferes with the metabolism of B vitamins, which can raise homocysteine levels. There are literally dozens of estrogen products, and I only listed the most commonly prescribed brands. If you are taking estrogen in the form of birth control pills or hormone replacement therapy, even if you don't see your particular estrogen drug on the list, you should assume that it works just like other estrogens and that you need to compensate for the loss of B vitamins by taking a B-complex supplement daily. The same recommendation holds true for other categories of drugs: even if you don't see your particular name brand, if you are taking a category of drug that can cause nutrient depletion, it is wise to take the appropriate supplements. For those who want a more complete list of drugs, I can recommend a wonderful reference book, *Drug-Induced Nutrient Depletion Handbook,* by Ross Pelton, James B. LaValle, Ernest B. Hawkins, and Daniel Krinsky. It's a virtual gold mine of information on the potential negative effects of drugs, and I think it should be in everyone's home library. See Appendix 3 for information on where to get it.

Drugs That Lower Coenzyme Q10

Having low Co-Q10 levels means more free radical damage to your brain, less energy for the brain to do its work, more tired, sluggish brain cells that don't learn, think, or remember as well as they should, and a greater susceptibility to neurological diseases.

Co-Q10 is made by every cell in the body but is particularly important for your brain. It is one of the few antioxidants that is fat soluble, which means it can get into the fatty brain cell membranes and protect them against free radical attack. Co-Q10 is also known as the "energizing antioxidant" because it is essential for the production of adenosine triphosphate (ATP), the fuel used by all cells, including brain cells, to do their work. Simply put, if you don't have enough Co-Q10, your brain cells cannot function effectively or efficiently. Your response time will be slower, your memory will grow dimmer, and you will not be able to achieve all that you should. My point is, most people need more Co-Q10, not less!

Production of Co-Q10 typically declines with age, just around the same time that you are beginning to experience those annoying "Why did I walk into this room?" memory problems and other signs of diminished brain function. (Like nodding out during a long meeting or being easily distracted.) The decline in Co-Q10 is a double whammy to your brain cells. At the same time they are running low on energy, they are also being assaulted by excess free radicals, which will further interfere with their ability to make energy. If your brain doesn't have enough energy, it cannot make enough neurotransmitters—the chemicals that keep you smart and alert—nor can it repair brain cells as they are damaged or "clean up" after free radicals. If you want to maintain optimal brain function, it's imperative that you protect and maintain your Co-Q10 levels. Low levels of Co-Q10 are linked to Parkinson's disease and may accelerate brain aging. I consider Co-Q10 to be so fundamental to brain health that I recommend it for everyone, including people following the low-risk Tier 1 program. (Co-Q10 supplementation: Tier 1, 30 milligrams daily; Tier 2, 60 milligrams daily; Tier 3, 200 milligrams daily.)

Many antidepressants, beta blockers prescribed to lower high blood pressure, antidiabetic drugs, and anticholesterol drugs inhibit the body's ability to make Co-Q10. Ironically, the conditions these drugs are meant to treat—depression, high blood pressure, and diabetes—put you at greater risk of heart disease. Having chronically low Co-Q10 levels will only make matters worse. It will also slowly and insidiously eat away at your brain cells, depriving them of energy and vital antioxidant protection.

Statin medicines to lower high cholesterol levels, including Lipitor, Mevacor, Pravachol, and Zocor, also deplete cells of Co-Q10. They are among the most widely prescribed drugs in the world. In my opinion, doctors are too quick to write prescriptions for these drugs, ignoring their potential downside. I'm not saying that these drugs are all bad; they're not. They can help prevent heart attacks, particularly for people who are resistant to making changes in diet and lifestyle. I do believe, however, that these drugs are terribly overused.

Statin drugs have another particularly nasty side effect: according to a recent study in the journal *Neurology,* patients using statin drugs had a stunning *1,600 percent increase* in peripheral neuropathy. Peripheral neuropathy is an often debilitating disease characterized by burning pain, tingling, numbness, and loss of sensation in the extremities. As a neurologist, I can tell you that peripheral neuropathy is extremely difficult to treat. Desperate patients turn to everything from epilepsy drugs to antidepressants to narcotic pain medications, often to no avail. With the adage "an ounce of prevention is worth a pound of cure" in mind, I believe that the medical profession should reconsider the widespread use of statin drugs. More physicians should follow the admonition of the Food and Drug Administration (FDA) in the *Physicians' Desk Reference* that these drugs should only be used "when dietary measures have failed." Before starting patients on statin drugs, they should be put on a serious nutritional intervention and exercise program, which can be highly effective in lowering cholesterol. I know that many of my colleagues would say that given a choice of taking a pill or watching what they eat, they would rather take the pill. My response is: give patients the real choice, and tell them the whole story. Tell them about the potential side effects of these so-called "wonder drugs." Let them know in full detail about the potential side effects of these drugs and why Co-Q10 is so vital for both brain and heart health. When patients are truly educated, they may have a change of heart and may be more willing to try dietary and lifestyle changes that are also proven methods of controlling blood lipids and reducing heart disease.

Below is a list of commonly used drugs that can deplete Co-Q10 levels. Read over this list carefully to see if you are taking any of these medications. Please note that given the hundreds of drugs on the market, I could not include every possible drug on this list. As you review the list, however, you will see that entire categories of drugs may deplete the body of Co-Q10 (for example, statin drugs to lower cholesterol deplete the body of Co-Q10, and so do many medicines that lower blood pressure). If you are taking a drug in any of the categories listed here that is not on this list, I strongly recommend that you take supplemental Co-Q10 anyway. It can only help your brain, and there are no known ill side effects. If you are taking a drug that depletes your body of Co-Q10 or think that you may be, please turn to page 46 to see what you need to do to compensate for the potential loss in Co-Q10.

Antidepressants

GENERIC NAME: Amitriptyline. BRAND NAME: Elavil

GENERIC NAME: Desipramine. BRAND NAME: Norpramin

GENERIC NAME: Doxepin. BRAND NAME: Sinequan

GENERIC NAME: Imipramine. BRAND NAME: Tofranil

GENERIC NAME: Nortriptyline. BRAND NAMES: Aventil, Pamelor

GENERIC NAME: Protriptyline. BRAND NAME: Vivactil

Antipsychotic Drugs

GENERIC NAME: Haloperidol. BRAND NAME: Haldol

Blood Pressure–Lowering Drugs

GENERIC NAME: Atenolol. BRAND NAME: Tenormin

GENERIC NAME: Bisoprolol. BRAND NAME: Zebeta

GENERIC NAME: Clonidine. BRAND NAME: Catapres

GENERIC NAME: Hydroclorothiazide (HCTZ); diuretic used alone or in combination with other medications to lower blood pressure. BRAND NAMES: Aldactazide, Capozide, Combipres, Dyazide, HydroDIURIL, Hyzaar, Lopressor-HCT, Lotensin HCT, Maxide, Microzide, Moduretic, Prinzide, Vasaretic, Zestoretic.

GENERIC NAME: Nadolol. BRAND NAME: Corgard

GENERIC NAME: Metoprolol. BRAND NAMES: Lopressor, Toprol

GENERIC NAME: Pindolol. BRAND NAME: Visken

GENERIC NAME: Propranolol. BRAND NAME: Inderal

Cholesterol-Lowering Drugs

GENERIC NAME: Atorvastatin. BRAND NAME: Lipitor

GENERIC NAME: Fluvastatin. BRAND NAME: Lescol

GENERIC NAME: Lovastatin. BRAND NAME: Mevacor

GENERIC NAME: Pravastatin. BRAND NAME: Pravachol

GENERIC NAME: Simvastatin. BRAND NAME: Zocor

Antidiabetic Drugs

GENERIC NAME: Glipizide. BRAND NAME: Glucotrol

GENERIC NAME: Glyburide. BRAND NAMES: DiaBeta, Glynase, Micronase

GENERIC NAME: Tolazemide. BRAND NAME: Tolinase

Solution

If you are taking a drug that depletes Co-Q10, you must take a Co-Q10 supplement daily.

Tier 1. If you are following the Tier 1 program, you are already taking 30 milligrams of Co-Q10. If you are taking a Co-Q10-depleting drug, take an additional 60 milligrams of Co-Q10 for a total of 90 milligrams daily (60 milligrams in the A.M. and 30 milligrams in the P.M.).

Tier 2. If you are on Tier 2, you are already taking 60 milligrams of Co-Q10 daily. If you are taking a Co-Q10-depleting drug, take an additional 60 milligrams of Co-Q10 for a total amount of 120 milligrams daily (60 milligrams in the A.M. and 60 milligrams in the P.M.)

Tier 3. If you are on Tier 3, you are already taking 200 milligrams of Co-Q10. If you are taking a Co-Q10-depleting drug, take an additional 100 milligrams of Co-Q10 supplement for a total amount of 300 milligrams daily (200 milligrams in the A.M. and 100 milligrams in the P.M.).

Drugs that Deplete B Vitamins

B vitamins are your best protection against elevated levels of homocysteine, an amino acid produced in the body that in excess, can increase your risk of mood disorders, poor mental performance, and Alzheimer's disease.

A wide range of drugs, from aspirin to estrogen to diuretics to stomach acid suppressors (such as Nexium and Prilosec) can interfere with the metabolism of one or more B vitamins, which can result in elevated levels of homocysteine. You who are worried about preserving your brain power, take note: medical research now clearly links poor performance on mental function tests to elevated homocysteine levels. Having too much of this amino acid floating around your bloodstream can cost you a few IQ points! B vitamins are also critical for mood and concentration, and a severe deficiency in B vitamins (especially vitamin B12) can result in symptoms such as severe confusion and mental fog at any age. Elevated levels of homocysteine and/or low levels of B vitamins can also increase the risk of depression, stroke, Alzheimer's disease, vascular dementia, heart disease, and even certain types of cancer. If you are using a drug that depletes your body of B vitamins, you must take supplemental B vitamins daily and demand to have your homocysteine levels checked periodically by your doctor.

You will notice that the most widely prescribed drug for Parkinson's disease, the generic drugs carbidopa and levodopa, two drugs marketed under the name Sinemet, can raise homocysteine levels! This probably explains why Parkinson's patients are at much higher risk of having a stroke. Sinemet is an effective drug for relieving the symptoms of Parkinson's, and I use it in my practice. Given the fact that elevated homocysteine is linked to poor cognitive performance in older people and increased inflammation, I am very careful to regularly monitor homocysteine levels for patients on this drug and to tailor my recommendations for B vitamin supplementation based on their results.

A word about NSAIDs: these drugs, such as the prescription drug Celebrex and the numerous over-the-counter brands of ibuprofen and naproxen, are commonly used to treat the aches and pains associated with arthritis. Some studies have shown that the long-term effects of anti-inflammatory drugs may reduce the risk of Alzheimer's disease and dementia. This is not surprising, considering the fact that inflammation is a major contributor to these problems, and anything that helps to reduce inflammation is helpful. The downside is that these drugs have the potential to raise homocysteine levels, which could aggravate neurological problems in susceptible people. Why risk it? Anyone who takes these drugs daily should also take supplementary B vitamins daily.

If you are taking a drug that destroys B vitamins, you must compensate by taking supplemental B vitamins. Review the list that follows to see if you are taking any drug that depletes B vitamins. Please note that, given the hundreds of drugs on the market, I could not include every possible drug. As you review the list, however, you will see that entire categories of drugs may deplete the body of B vitamins. For example, even if you don't see your particular estrogen drug on the list, assume that it works just like other estrogens and that you need to compensate for the loss of B vitamins by taking B complex daily. The same recommendation holds true for other categories of drugs—even if you don't see your particular name brand, if you are taking a category of drug that can cause nutrient depletion, it is wise to take the appropriate supplements.

Pain Relievers

GENERIC NAME: Aspirin (acetylsalicylic acid). BRAND NAMES: Pure aspirin is sold under numerous brand names, including many private store labels. Percodan and Empirin are combination aspirin and codeine products.

Antibiotics

GENERIC NAME: Trimethoprim: antibiotic frequently prescribed for chronic urinary tract infections. BRAND NAMES: Bactrim, Septra

Antacid and Stomach Acid Suppressors

GENERIC NAME: Cimetidine. BRAND NAME: Tagamet

GENERIC NAME: Famotidine. BRAND NAME: Pepcid

GENERIC NAME: Ranitidine. BRAND NAMES: Zantac, Zantac 75

GENERIC NAME: Lansoprazole. BRAND NAME: Prevacid

GENERIC NAME: Nizatidine. BRAND NAME: Axid

GENERIC NAME: Omeprazole. BRAND NAME: Prilosec

Antidiabetic Drugs

GENERIC NAME: Metformin. BRAND NAME: Glucophage

Asthma Drugs

GENERIC NAME: Beclomethasone (oral inhaler). BRAND NAME: Vanceril

GENERIC NAME: Budesonide (oral inhaler). BRAND NAME: Pulmacort

GENERIC NAME: Budesonide (nasal inhaler). BRAND NAME: Rhinocort

GENERIC NAME: Flunisolide (nasal inhaler). BRAND NAME: Nasalide

GENERIC NAME: Flunisolide (oral inhaler). BRAND NAME: Aerobid

GENERIC NAME: Fluticasone (oral inhaler). BRAND NAME: Flovent

GENERIC NAME: Mometasone (nasal inhaler). BRAND NAME: Nasonex

GENERIC NAME: Triamcinolone (oral inhaler). BRAND NAME: Azmacort

GENERIC NAME: Theophylline. BRAND NAME: Aerolate

Blood Pressure–Lowering Drugs

GENERIC NAME: Bumetanide. BRAND NAME: Bumex

GENERIC NAME: Hydroclorothiazide (HCTZ), used alone or in combination with other drugs. BRAND NAMES: Aldactazide, Capozide, Combipres, Dyazide, HydroDIURIL, Hyzaar, Lopressor-HCT, Lotensin HCT, Maxide, Microzide, Moduretic, Prinzide, Vaseretic, Zestoretic.

GENERIC NAME: Triamterene. BRAND NAMES: Dyrenium, or in combination with other drugs in Maxide and Dyazide.

GENERIC NAME: Furosemide. BRAND NAME: Lasix

GENERIC NAME: Hydralazine. BRAND NAME: Apresoline

GENERIC NAME: Torsemide. BRAND NAME: Demadex

Anticonvulsant Drugs

GENERIC NAME: Carbamazepine. BRAND NAME: Tegretol

GENERIC NAME: Ethosuximide. BRAND NAME: Zarontin

GENERIC NAME: Fosphenytoin. BRAND NAME: Cerebyx

GENERIC NAME: Mephobarbital. BRAND NAME: Mebaral

GENERIC NAME: Phenobarbital. BRAND NAME: Phenobarbital

GENERIC NAME: Phenytoin. BRAND NAME: Dilantin

GENERIC NAME: Primidone. BRAND NAME: Mysoline

GENERIC NAME: Valproic Acid. BRAND NAME: Depakote, Depakene

Cholesterol-Lowering Drugs

GENERIC NAME: Cholestyramine. BRAND NAME: Colestid

Estrogens

There are dozens of different brands of estrogen products, and I have only listed the major brands here. If you taking estrogen in the form of birth control pills or hormone replacement therapy, even if you don't see your particular estrogen drug on the list, assume that it works just like other estrogen and that you need to compensate for the loss of B vitamins by taking additional B vitamins.

GENERIC NAME: Estrogens (with or without progesterone) as contraceptives, oral or skin patch. BRAND NAMES: Ortho-Novum, OrthoTri-Cyclen, Ovral, Ovcon, Demulen, Loestrin (any estrogen-containing oral contraceptive.)

GENERIC NAME: Estrogens (with or without progesterone) for hormone replacement therapy for menopause or hysterectomy; oral, skin patch, or cream. BRAND NAMES: Premarin, Prempro, Activela, Combinpatch, Estrotest, any hormone replacement product containing estrogen.

Estrogen Substitutes for Osteoporosis

GENERIC NAME: Raloxifene. BRAND NAME: Evista

Anti-Parkinson's Drugs

GENERIC NAME: Carbidopa and levodopa. BRAND NAME: Sinemet

Nonsteroidal Antiinflammatory Drugs (NSAIDs)

GENERIC NAME: Celecoxib. BRAND NAME: Celebrex

GENERIC NAME: Ibuprofen. BRAND NAME: Advil, Bayer Select, Motrin, Midol, etc.

GENERIC NAME: Indomethacin. BRAND NAME: Indocin

GENERIC NAME: Naproxen. BRAND NAMES: Naprosyn, Aleve, etc.

Corticosteroids: Anti-inflammatory Drugs

Used to treat asthma, arthritis, allergy, and pain.

GENERIC NAME: Methyl Prednisolone. BRAND NAME: Medrol

GENERIC NAME: Prednisone. BRAND NAMES: Deltasone, Orasone

Solution

If you are taking a drug that depletes your body of B vitamins, you must compensate for the loss by taking a B vitamin supplement. Everyone on the Better Brain program is taking some B vitamin supplements. If you are taking a B-depleting drug, you may need to increase your dose. It is also essential for you to have your homocysteine checked after starting the new drug to make sure that it isn't rising to an unhealthy level. Specific information follows for people following the different tiers of the Better Brain program.

Tier 1. If you are following the Tier 1 program, you are a taking a basic B-complex supplement containing at least:

B1 (thiamine)	50 mg
B3 (niacin as niacinamide)	50 mg
B6 (pyridoxine)	50 mg
Folic acid	400 mcg
B12 (cobalamin)	500 mcg
Look for a B-complex supplement containing several B vitamins in one capsule.	

1. Within three months of starting a B-depleting drug, have your homocysteine level checked by your physician. It's a simple blood test that can be done right at your doctor's office and sent to a laboratory for analysis. (See chapter 10 for more information on this vital test.) *A homocysteine level of over 9 micromoles per liter dramatically increases your risk of neurological problems and should therefore be treated.*

2. If your homocysteine level is normal, keep doing what you are doing. It's working. Have your homocysteine level checked every year as part of your annual physical.

3. If your homocysteine level is elevated after starting the drug, continue taking your B complex but increase your dose of the following supplements:

Increase folic acid to 1.2 mg daily; 800 mcg in the A.M. and 400 mcg in the P.M.

Increase B12 to 1,000 mcg once daily; 500 mcg in the A.M. and 500 mcg in the P.M.

4. Recheck your homocysteine level in two months. If it is still elevated, add to your regimen another supplement, trimethylglycine (TMG). A naturally occurring substance found in plants and animals, TMG can help convert harmful homocysteine back into methionine, which is beneficial to the body. You can find it in any health food store and most pharmacies. Take 1,000 milligrams of TMG daily.

5. Recheck your homocysteine level in two months. If it is not normalized, talk to your physician about getting shots of vitamin B12.

6. Once your homocysteine level is normalized, have it checked every six months.

Tier 2. If you are following the Tier 2 program, you are taking

B1 (thiamine)	50 mg
B3 (niacin as niacinamide)	50 mg
B6 (pyridoxine)	50 mg
Folic acid	400 mcg (twice daily)
B12 (cobalamin)	500 mcg (twice daily)
Look for a B-complex supplement containing several B vitamins in one capsule or pill.	

1. Within three months after starting on the B-depleting drug, have your homocysteine level checked by your physician. It's a simple

blood test that can be done right at your doctor's office and sent to a laboratory for analysis. (See chapter 10 for more information on this vital test.) *A homocysteine level of over 9 micromoles per liter dramatically increases your risk of neurological problems and should therefore be treated.*

2. If your homocysteine is normal after starting the drug, keep doing what you are doing. It's working. Have your homocysteine level rechecked every six months.

3. If your homocysteine levels are elevated after starting the drug, continue taking your B complex but add another supplement, trimethylglycine (TMG). A naturally occurring substance found in plants and animals, TMG can help convert harmful homocysteine back into methionine, which is beneficial to the body. You can find it in any health food store and most pharmacies. Take 1,000 milligrams of TMG daily.

4. Recheck your homocysteine level in two months. If it is not normalized, talk to your physician about getting shots of vitamin B12.

5. Once your homocysteine level is normalized, have it checked every six months.

Tier 3. If you are following the Tier 3 program, you are taking:

B1 (thiamine)	50 mg
B3 (niacin as niacinamide)	50 mg
B6 (pyridoxine)	50 mg
Folic acid	400 mcg (twice daily)
B12 (cobalamin)	500 mcg (twice daily)

1. Have your homocysteine level checked by your physician within three months of starting the new drug. It's a simple blood test that can be done right at your doctor's office and sent to a laboratory for analysis. (See chapter 10 for more information on this vital test.) *A homocysteine level of over 9 micromoles per liter dramatically increases your risk of neurological problems and should therefore be treated.*

2. If your homocysteine is normal after starting the drug, keep doing what you are doing. It's working.

3. If your homocysteine levels are elevated after starting the drug, continue taking your B complex, but add another supplement to your regimen, trimethyglycine (TMG). A naturally occurring substance found in plants and animals, TMG can help convert harmful homocysteine back into methionine, which is beneficial to the body. You can find it in any health food store and most pharmacies. Take 1,000 milligrams of TMG daily.

4. Recheck your homocysteine level in two months. If it is not normalized, talk to your physician about getting shots of vitamin B12.

5. Once your homocysteine level is normalized, have it checked every six months.

Drugs that Lower Glutathione

Acetaminophen, a nonaspirin pain reliever that is sold under many different brand names, can reduce levels of glutathione, a critically important brain antioxidant. Glutathione is essential for removing toxins from the body that can promote free radicals and create inflammation in the brain. Every day we are exposed to thousands of toxins, in the air we breathe, the food we eat, the drugs we take, and the other chemicals we ingest or use in our homes, knowingly and unknowingly. (For more information on toxins and your brain, see chapter 8.) It's the liver's job, with the help of glutathione, to render these chemicals harmless. Without adequate levels of glutathione, the liver, the primary detoxifying organ of the body, cannot function properly, and the result will be a buildup of toxins throughout the body and in the brain.

This is bad news for your brain. If your brain is being assaulted by free radicals, and there is not enough glutathione on hand to defend it, it will degenerate all the faster. You really will be "losing it," maybe not overnight but certainly over time. The free radical–inflammation cycle is responsible for the mild memory and performance problems that start in midlife and lead to serious brain disorders decades later. You need to stop it, and glutathione is an important weapon in the lifelong fight against free radicals.

Low levels of glutathione are also associated with chronic illness and premature death. As we age, we naturally produce less glutathione, which significantly weakens our antioxidant defenses. One of the primary goals of the Better Brain program is to keep your glutathione levels high. The occasional use of acetaminophen is not a problem (once a week or so), but lately it has been marketed as a drug you can take daily for the chronic pain of arthritis. At least one study has shown an increased risk of Alzheimer's disease in people who have routinely used acetaminophen for two or more years, and I don't recommend using this drug every day.

If glutathione is critical to our health, why don't we just replenish it with glutathione supplements? Taken orally, glutathione is not well absorbed by the body. (That's why I give intravenous glutathione to my Parkinson's patients, who need it the most. See chapter 15.) Fortunately, there are a few special supplements that can raise glutathione levels, including N-acetyl-cysteine, or NAC (see chapter 6). These antioxidants give the body a glutathione boost—not as strong as giving intravenous glutathione but enough to make a difference over time. If you are following the Tier 1 supplement regimen, you are already taking 400 milligrams of NAC daily; if you are following Tier 2, you are already taking 800 milligrams daily. If you are using a drug that depletes glutathione, you should increase your dose of NAC. Of course, it's far better if you protect your body's own supply of glutathione by reducing exposure to toxins and avoiding drugs that deplete you of glutathione.

Review the list that follows to see if you are taking any drug that depletes glutathione. Please note that, given the hundreds of drugs on the market, I could not include every possible drug. Don't assume that if a product is not on this list it doesn't contain acetaminophen. Given the widespread use of acetaminophen, any time you purchase a nonaspirin pain reliever or an over-the-counter cold or sinus treatment, you must check the ingredient label to see if it contains acetaminophen; that is the only way you can know for sure.

Nonaspirin Pain Relievers

GENERIC NAME: Acetaminophen. BRAND NAMES: Panadol, Tylenol, Tylenol Arthritis Pain, Aspirin-Free Anacin, plus countless house brands. Acetaminophen may also be added to over-the-counter cold medicines and codeine-containing pain relievers.

Solution

Try to avoid using these drugs daily. If you are routinely taking a glutathione-depleting drug more than twice a week, please take an N-acetyl-cysteine (NAC) supplement to boost glutathione levels. Specific information follows for people following the different tiers of the Better Brain program.

Tier 1. If you are following the Tier 1 program, you are currently not taking an NAC supplement. If you are using a glutathione-depleting drug, take 400 milligrams of NAC once daily in the A.M.

Tier 2. If you are following the Tier 2 program, you are currently taking 400 milligrams of NAC daily. Take an additional 400 milligrams of NAC daily for a total of 800 milligrams daily. Divide your dose into 400 milligrams in the A.M. and 400 milligrams in the P.M.

Tier 3. If you are following the Tier 3 program, you are already taking 800 milligrams of NAC daily. Take an additional 400 milligrams of NAC daily, for a total of 1,200 milligrams Take 800 milligrams in the A.M. and 400 milligrams in the P.M.

Avoid drinking alcohol when you take acetaminophen. Alcohol also depletes the liver of glutathione, and the combination can be deadly.

Drugs That Contain Aluminum

Aluminum is a metal that causes the formation of brain-damaging free radicals and inflammation. There are abnormally high accumulations of aluminum in the brains of people with Alzheimer's disease. Some brands of antacids contain high amounts of aluminum, and this worries me because many people pop antacids as if they are candy. Although the degree to which aluminum may be a causative factor in Alzheimer's disease is controversial, some studies show a direct correlation between aluminum content in the water supply and the incidence of Alzheimer's disease. Researchers at the University of Toronto found an astounding 250 percent increased risk of Alzheimer's disease in individuals drinking municipal water high in aluminum for more than 10 years. Many of us are already exposed to high levels of aluminum in everyday life, not just in water or antacids, but in common products such as cookware, foil, and even underarm deodorants. (See chapter 6 for more information.) Although I believe that people should make a concerted effort to use aluminum-free products whenever possible, I'm not concerned about the occasional use of antacids. If you use them once or twice a month, I don't think you need to worry. Many people, however, pop antacid pills throughout the day, and these people are at risk of accumulating high levels of aluminum. If you use antacids daily, I recommend that you look for an aluminum-free product. Even better, why not try to deal with your chronic indigestion, the real source of your problem? Changing your eating habits—eliminating processed and/or fried foods and sticking to a wholesome diet—is often all it takes to restore healthy digestion. (For more information on healthy eating, see chapter 5.) Another concern about antacids is that, like the prescription products, they can raise homocysteine by depleting the body of B vitamins.

Read over the list of aluminum-containing drugs to see if you are taking any of them. Given the scores of different antacid products on the market and new ones being brought to market every day, there may be some aluminum-containing antacids that are not on this list. Don't assume that if a product is not on this list, it's fine. The only way you can know for sure is to check the ingredients label for aluminum.

Antacids

GENERIC NAME: aluminum hydroxide (with magnesium carbonate, or magnesium hydroxide, magnesium hydroxide and simethicone, or magnesium trisilicate). DRUG TYPE: antacid. BRAND NAMES: Gaviscon, Aludrox, Di-Gel, Gelusil, Maalox, Magalox, Mylanta.

Solution

Use these drugs rarely, if at all. By *rarely* I mean no more than twice a month. If you need to use antacids more frequently, switch to drugs containing calcium carbonate instead of aluminum, such as Tums and Rolaids.

Tools for a
Better Brain

Building a Better Brain Through Nutrition

I AM OFTEN ASKED what I think is the single most important thing you can do to keep your brain functioning at its peak and prevent brain aging. The answer is easy. If you want to perform at the highest levels and maintain a lifetime of optimal brain health, you must be vigilant about what you put on your plate. It's as simple as that. Nutrition is *the* most important tool for staying mentally and physically fit, and it is by far the most underutilized tool.

The Better Brain meal plan will give you the tools you need to make good choices so that you can take advantage of the power of good nutrition. The meal plan lays out for you four weeks' worth of meals and snacks incorporating the principles of the Better Brain program. These menus are meant to be used as a guideline; please adapt them to suit your personal likes and dislikes. Please feel free to create your own meals from the food lists also provided here.

From the perspective of a neurologist, the standard American diet is a nightmare. It is loaded with poor-quality fat that can make your brain cells sluggish but is scarce in healthy fat that can keep your brain cells flexible and "smart." It is packed with highly processed, nutrient-deficient food that is laden with sugar and chemical additives that practically invite free radicals and inflammation to invade your brain. If I were to design a diet with the sole purpose of creating an epidemic of poor brain function, accelerated brain aging, mood disorders, and other neurological problems, it would be the one that most Americans are already following.

Fortunately, food is powerful medicine. Simple changes in your diet can protect your body and your brain against aging, debility, and disease. You have a choice. You don't have to follow the standard American diet and continue to walk around in a forgetful fog! Get the fattening, starchy junk foods off your plate (refined, processed grains such as white bread, white rice, chips, white pasta, sweetened cereals) and replace them with real food such as unprocessed, whole grains loaded with brain-boosting B vitamins and antioxidant-rich fruits and vegetables. (Later I give a complete list of "brain-building" foods that will improve mental performance and a list of "brain busters" that you should eat rarely, if at all.) Eliminate the unnecessary sources of sugar (such as soda and other sweetened beverages and snack foods) that are adding on those extra pounds and wildly accelerating the formation of free radicals. Buy organic produce and meat that are free of added chemicals such as pesticides, growth hormones, and antibiotics that can spread inflammation throughout your body. All of these changes can add up to significant health benefits for body and mind alike.

Change Your Fat, Change Your Brain

The most important nutrient for your brain is fat. The brain, the most metabolically active organ in the body, is constantly replenishing and repairing itself. Much of the food we eat not only ends up being used by the brain for energy but ends up *in* the brain itself. The brain needs fat more than any other nutrient, because the brain is made of fat. The problem is that if we feed it unhealthy fats, we wind up with an unhealthy brain

There are four primary categories of fat in the foods we eat: monoun-saturated fat, saturated fat, polyunsaturated fat, and trans-fatty acids. The first of these three occur naturally in food. The fourth group, trans-fatty acids, is a creation of modern chemical processing and has slowly and insidiously crept into the food supply, and into your brain.

Monounsaturated Fat

This fat is a brain-friendly fat. It is found in common cooking oils, in-cluding olive oil, canola oil and some forms of safflower oil, nuts, and av-ocados. Naturally high in antioxidants, monounsaturated fats are subject to less oxidative damage than other types of fat. That means that when they become incorporated into your brain cells, they are less vulnerable to free radical attack, which means they are less prone to damage.

Saturated Fat

Saturated fat doesn't make high-energy, agile brain cells; it makes slug-gish brain cells. In limited amounts it's fine (about 10 percent of your daily caloric intake is okay), but most Americans eat about three times that amount of saturated fat every day. Saturated fat is found primarily in foods of animal origin, including beef, lamb, pork, chicken, eggs, and whole-fat dairy foods. These fats are more prone to oxidative damage than monounsaturated fats, which can increase the risk of free radical damage to your cells. A diet high in saturated fat can raise the level of homocysteine, the amino acid that in excess is toxic to your brain. At any age, high levels of homocysteine can cause memory problems and mood disorders and may increase your risk of developing Alzheimer's disease.

Polyunsaturated Fat

Some forms of polyunsaturated fat are great for your brain; the problem is, we don't eat enough of them. Polyunsaturated fats include the all-important *essential fatty acids,* fats that cannot be made by the body but must be obtained through food. Essential fatty acids are critical for a well-functioning brain, but some are more important for your brain than others. There are two types of essential fatty acids: omega 3 fatty acids and omega 6 fatty acids. Omega 3 fatty acids are found in cold water fatty fish, deep green vegetables (purslane, a leafy green vegetable that has become popular lately, is an excellent source), and some grains and seeds (pumpkin seeds). *Most people do not get enough omega 3 fatty acids in their diet.*

What is the best fat for your brain? In the body, omega 3 fatty acids from food are broken down into two other fatty acids, eicosapentanoic acid (EPA) and docosahexanenoic acid (DHA). Healthy, well-functioning brains contain high amounts of DHA, which provides the perfect raw material for well-functioning cell membranes. But maintaining the optimal amount of DHA for the brain can be a problem. Although the body has no trouble making enough EPA, many people are unable to produce adequate amounts of DHA on their own. Excess consumption of bad fats (saturated fat and trans-fatty acids) and alcohol can interfere with the conversion of omega 3s to DHA. What happens if you don't make enough DHA? Low levels of DHA in adults are associated with decreased cognitive function, depression, moodiness, irritability, slow response time, and Alzheimer's disease.

Omega 6 fatty acids are found in cooking oils, nuts, and most seeds and cereals. Vegetable oils (corn oil, peanut oil, sunflower oil, and margarine) are the primary sources of omega 6 fatty acids in the diet and unfortunately, are not the best fats for your brain. Commercially processed oils and margarine are converted into trans-fatty acids in the body, which promote free radical production and inflammation. Although people tend to get enough omega 6 fatty acids from food, they may be deficient in one type of omega 6 fatty acid, gammalinoleic acid (GLA). Good sources of GLA include avocados, walnuts, seeds, and borage oil, an oil made from the borage plant that is sold in capsules as a nutritional supplement. As GLA has powerful anti-inflammatory properties, I often prescribe GLA supplements to people with multiple sclerosis and Alzheimer's disease, diseases characterized by inflammation.

Trans-Fatty Acids

These fats make rigid, tough, slow brain cells. *Get these fats off your plate.* Many brands of margarine and polyunsaturated oils undergo a chemical modification to extend their shelf life and make them easier to use in baking. This process creates trans-fatty acids, a synthetic fat that is unlike any fat found in nature. Trans fats are the primary fats found in processed baked goods and fried foods. They are so prevalent in the food supply that unless a label for a processed food specifically says "no trans-fatty acids" you must assume that it may contain some. At one time, all margarine contained trans-fatty acids; now there are some brands on the market that do not.

What's wrong with trans-fatty acids? Like any other dietary fat, trans-fatty acids become incorporated into your cell membranes. Unlike healthier fats, trans-fatty acids can make your cell membranes hard and rigid. This destroys the cell's ability to make energy, get adequate nutrition, or communicate with other cells. Trans-fatty acids will interfere with your brain's ability to perform well. *A brain filled with trans-fatty acids is going to age faster and become progressively less functional.* These fats are also linked to an increased risk of diabetes and heart disease, two ailments associated with an increased risk of dementia and depression.

Caution: Perhaps the most dangerous aspect of trans-fatty acids is that they crowd out healthier fats. Even if you eat lots of good fat along with trans-fatty acids and take an essential fatty acid supplement, the trans-fatty acids are the ones that will end up in your cell membranes.

Eat More Good Fat

Eating a diet rich in omega 3 fatty acids is a good first step toward increasing your level of DHA. In the suggested menus in this chapter, I show you how to add foods rich in omega 3 fatty acids to your daily diet. But be aware that *simply adding more omega 3 fatty acids to your diet will not make a better brain.* At the same time you are increasing your intake of omega 3 fatty acids, you must *decrease your intake of trans-fatty acids and saturated fats.* Even if you increase your intake of omega 3 foods, however, I still recommend that everyone take at least 300 milligrams of DHA in supplement form daily. This way you will ensure that you are getting ample amounts of DHA even if your body cannot produce it on its own from food. (If you are following Tier 1 or Tier 2, you will be taking 300 milligrams of DHA daily. If you are following Tier 3, you will be taking 600 milligrams of DHA daily.)

What to Do

The steps that follow show you how to get the bad fat out and the good fat in.

Step 1: Cut Out the Trans-Fatty Acids

Don't buy products that contain "partially hydrogenated vegetable oil" or partially hydrogenated vegetable shortening. These include most (and I mean nearly all) processed, refined foods such as commercial bread, crackers, frozen waffles, cookies, and snack foods. Don't assume that a so-called healthy food, such as whole-grain bread, does not contain hydrogenated oils. You must read the ingredients label to be sure.

Don't deep-fry your food. Any fat heated at a high temperature (with the oil bubbling hot) will result in trans-fatty acids. Instead bake, sauté (with a *low* flame), steam, or grill.

Don't eat fried food—this includes french fries, donuts, and most chips (corn, cheese, and potato). Baked chips may contain hydrogenated oils, so read the labels before eating these foods.

Trans-fatty acids pop up in the least likely places, from so-called health food cereals to packaged baked goods to french fries. *Check food labels for trans fat content.* If a food contains trans-fatty acid, don't buy it.

Step 2: Reduce Your Intake of Saturated Fat

Meat and full-fat dairy products are the major sources of saturated fat in the American diet. They are also the major source of protein in our diet. Protein is important for cell repair and maintenance, and you need adequate protein to keep your brain functioning at its peak. But you don't have to choose proteins that contain bad fat. For example, not all cuts of beef are high in saturated fat. If you choose to eat beef, stick to the leanest cuts of beef. Grass-fed, free-range beef is the best because it contains the least fat and is raised on organically grown food (without pesticides). You can buy it in many health food supermarkets and by mail order (see Appendix 3). Game meats (venison, buffalo) are growing in popularity in the United States and are extremely lean. You don't have to give up dairy, but use reduced-fat or no-fat dairy products, and please stick to organic products.

Get some of your protein from plant sources. Soy-based foods, such as tofu (bean curd) and tempeh (fermented soybean patty) are low in saturated fat and are good alternatives to meat. Try to have one or two soy meals a week. Soy contains beneficial antioxidants, called isoflavones, that protect against free radical attack. Please don't gorge on soy foods. Soy contains hormonelike compounds that in some studies appear to protect against hormone-sensitive cancers such as breast cancer and prostate cancer but in other studies have been shown to stimulate the growth of some cancers. To be on the safe side, don't exceed two to three soy meals a week. Other legumes (lentils, chickpeas, kidney beans) are also good sources of protein when combined with a starch (such as rice or corn.) The amino acids in grains and corn provide the amino acids that are missing from the legumes. Legumes are low in saturated fat, which is good because as you eat more omega 3 fatty acids, you don't want them crowded out of your brain cells by the more aggressive saturated fats.

Step 3: Increase Your Intake of Omega 3 Fatty Acids

Fatty fish is a terrific source of omega 3 fatty acids but may contain a fair amount of mercury, a toxic metal that is particularly dangerous for your brain. There are some exceptions: Alaskan sockeye salmon caught in the wild is rich in omega 3s and is one of the cleanest fish in terms of toxins. Try to eat Alaskan salmon at least twice a week. Tilapia, a white fish from Mediterranean waters, is also extremely low in mercury and is an excellent food choice. (See the lists later in this chapter.) Avoid farm-raised fish because they are typically fed hormones, antibiotics, and other "toxins" you don't need in your body. If you buy fresh fish at a fish store or a supermarket, be sure to ask the manager of the department whether the fish is farm raised and where the fish is from. If you buy canned fish, unless the label specifically says "Alaskan sockeye salmon," which means it is from Alaskan waters, you may have to call the manufacturer for more details.

Fresh greens provide some omega 3 fatty acids. One green in particular, purslane, is a great source of good fats and is sold in many health food stores and some supermarkets. It's great sautéed as a side dish (I have provided a recipe).

Snack on *walnuts and pumpkin seeds*. They are rich in omega 3 fatty acids and GLA. Walnut and pumpkin seed oil can be used in salad dressing.

Sprinkle *flaxseed* in your cereal, yogurt, or salad. It is also an excellent source of omega 3s. Ground flaxseed is sold at most health food stores; it has a mild, nutty taste. Flaxseed oil can be used in salad dressing. (See Appendix 2.)

Some brands of *eggs* contain higher amounts of omega 3 fatty acids and less saturated fat because they are from chickens that are fed special vegetarian diets. They are hormone and antibiotic free. You can find them at most supermarkets and health food stores. It is well worth the few extra cents you pay to get a better quality egg.

As a rule, most *processed baked goods* are hidden sources of trans-fatty acids, but there are a handful of brands that don't have trans-fatty acids, and are enriched with omega 3s. For example, Vons frozen waffles are enriched with omega 3 fatty acids and they taste great. Vons is sold in most health food stores and many supermarkets. There are also several brands of cereal that are enriched with omega 3s. Vote with your pocketbook and support these products.

Add Antioxidants to Your Diet

The fat in your brain—even good fat—is vulnerable to free radical attack, which is why you need to increase your intake of antioxidant foods, primarily colorful fruits and vegetables. I recommend that you try to eat at least six servings of vegetables and two servings of fruit every day. (One serving is the equivalent of ½ cup cooked vegetables, 1 cup chopped raw fruit or vegetable, or 1 medium-sized fruit, such as an apple or pear.) In addition to vitamins, fruits and vegetables contain hundreds of beneficial photochemicals that offer many unique benefits. Many photochemicals are natural antioxidants that help protect fruits and vegetables from free radical attack and fortify them against bacteria, fungi, and parasites. They do the same thing for humans. Moreover, photochemicals work synergistically, meaning that in order to get their full benefit, you need to get them in food with other photochemicals.

Many of the important photochemicals are in the pigments of plants. For example, orange and yellow fruits and vegetables have completely different photochemicals from those in dark green leafy vegetables. Foods with the most intense colors have the most antioxidants. Eating a wide assortment of differently colored fruits and vegetables every day is the best way to ensure that you are taking full advantage of nature's healing pharmacy.

The following steps will help you to get more antioxidants on your plate.

Step 1: Eat a Mixed Green Salad Every Day

Use several different types of lettuce and greens, such as arugula, endive, radicchio (red lettuce), and Romaine lettuce. It makes a tastier, more interesting salad, and it contains a broad spectrum of antioxidants. To save time, you can wash your salad ahead of time, wrap it in paper towels, and store it for up to two days in the produce section of your refrigerator.

Step 2: Eat More Cruciferous Vegetables

Cruciferous vegetables include broccoli, beet greens, watercress, collard greens, mustard greens, Swiss chard, kale, cabbage, brussels sprouts, and cauliflower. They are rich in vitamin E, folic acid, and the photochemical antioxidants lutein and zeaxanthin, which have been shown to improve mental performance in animals. Steam or lightly sauté vegetables to preserve their antioxidant content.

Get the Poisons Off Your Plate

The downside of fruits and vegetables is that they are often treated with pesticides, powerful neurotoxins that stay in your body and can cause significant damage to your brain and nervous system. They do their harm by increasing the production of free radicals in the body and promoting inflammation.

The vast quantity of artificial chemicals consumed in the diet poses a special burden on the body's detoxification system, which has the job of detoxifying potential poisons before they can harm healthy cells and tissues. Taking antioxidant supplements can help by keeping free radicals in check and reducing inflammation, but antioxidants are no match for the sheer numbers of chemicals bombarding the body daily. I understand that we can't eliminate all toxins from our lives, but we can make a concerted effort to reduce our exposure to some of the main offenders.

Step 1: Buy Organic Produce Whenever Possible

Organic produce is grown without pesticides or other chemical additives. Recently, the United States Department of Agriculture (USDA) has begun certifying organic products. Those that meet the grade can carry the USDA seal. Organic produce may be somewhat more expensive, but you're actually getting more for your money. Several studies have shown that organically grown produce retains significantly more vitamins and minerals than conventionally grown produce. Remember, just because produce is labeled organic, it doesn't mean that it doesn't need to be washed! Organic produce may be prone to fungal or bacterial infection and should be carefully washed in water. (If you live in an area where you don't have access to organic products, check Appendix 3 for mail order options.)

Step 2: Reduce Your Exposure to Pesticides

If organic produce isn't available, wash your produce in a mild soap solution and then rinse thoroughly in cool water. Discard the outer leaves of cabbage and lettuce where pesticides are more likely to concentrate. If you don't use organic produce, please peel your apples, pears, peaches and cucumbers (if waxed) before eating. This reduces fiber and vitamin content but it also significantly reduces pesticide exposure.

Step 2: Buy Whole, Natural Foods

Read food labels. If a product contains a lot of chemicals you've never heard of, don't buy it. (You should be reading labels anyway to avoid trans-fatty acids.)

Find a supermarket with a good selection of preservative-free, organic food, or shop in a health food store. I recommend buying organic products or products made with organic ingredients whenever possible. In my house, we only use organic dairy products derived from naturally raised, antibiotic-free cows and bread made from organic wheat. There are three types of organic labels. Products that are entirely organic will have a "100 percent organic" label. Products that contain at least 95 percent organic ingredients will have a USDA organic stamp. Products that have at least 70 percent organic ingredients will not have a USDA seal but can say that they are "made with organic ingredients." Keep in mind that the other 30 percent may be high in toxic pesticides.

What to Drink

Unfortunately, many municipal water systems contain high amounts of chlorine, a known brain toxin that should be filtered out of your drinking water. Moreover, tap water may also contain insecticides and other toxins. (See chapter 8.) Drink chemical-free, bottled water or install a water purification system in your home.

Coffee or tea is fine as long as you are not caffeine sensitive. Water-decaffeinated coffee is an option if you want to cut back on caffeine, but keep in mind that it may contain up to 50 percent of the caffeine found in regular coffee. Be sure that the coffee is water decaffeinated, which means it has undergone a chemical-free process to reduce caffeine content. In some cases, chemical solvents are used to decrease caffeine content in coffee, which can leave toxic residue. If you want to eliminate caffeine altogether, herbal tea is your best alternative.

Don't drink soda or other sweetened beverages. Please avoid diet soda—as discussed earlier, it contains artificial sweeteners that can overexcite brain cells, and damage them (so-called excitotoxins). Juice can add a surprising amount of sugar to your diet, and a high-sugar diet has been linked to memory problems. Use juice as a treat. I prefer freshly made vegetable juices, which are low in sugar and high in antioxidants.

Wine, beer, or any other alcoholic beverage may be beneficial to your brain. Alcohol helps lower blood pressure, which improves the flow of blood to the brain. It also helps to relieve stress. Red wine in particular contains powerful antioxidants. More than two alcoholic drinks daily, however, increases your risk of brain degeneration. (*One drink* means 4 ounces of wine, one 1-ounce shot glass of spirits, or one 8-ounce glass of beer.) Remember, alcohol must be detoxified by the liver, and this can use up the body's supply of glutathione, the antioxidant essential for detoxifying free radicals in the brain and everywhere else. Too much alcohol can make glutathione scarce, which will leave you at the mercy of free radicals. Alcohol is not for everyone. Don't drink if you have a problem with alcohol dependency, or are taking any medication that should not be combined with alcohol, or have a physical or mental condition that can be aggravated by alcohol, such as depression or liver disease.

Foods That Can Make You Smarter

Berries. You can teach old rats new tricks, as long as you feed them berries. Aging rats fed a blueberry-supplemented diet for four months scored as well as younger rats in their ability to recognize objects after an hour. Older rats fed a normal diet were clueless! Blueberries (and blackberries) contain anthocyanins, powerful antioxidants that help protect small blood vessels from free radical damage and may improve the blood flow within the brain. How can you get more berries in your life? Put ½ cup of blueberries or blackberries in plain low-fat or no-fat yogurt, sprinkle them on hot whole-grain cereal, put them in fruit salad, or use them to flavor a protein shake. If berries are out of season, use frozen, organic berries, which are sold in many health food stores. Try to eat two to three ½-cup servings of berries a week.

Omega 3–enriched eggs. These eggs are produced from hens fed a vegetarian, organic diet that is high in beneficial fatty acids and vitamin E, a brain-protective antioxidant, which makes their eggs healthier. Eggs are also a great source of B vitamins, which help keep homocysteine levels down. At one time, people were discouraged from eating eggs because the yolks are high in cholesterol (200 milligrams on average), which is believed to increase the risk of heart disease. We now know that cholesterol in food has little effect on cholesterol levels in the body, and there is no evidence that eating eggs increases cholesterol levels. So do eat eggs, but make sure that they are omega 3–enriched eggs from free-range chickens. You can eat up to between six to eight eggs a week.

Purslane. Purslane is actually a weed that is finding its way into salads and is being served as a side dish. It has a mild, nutty flavor and a crunchy texture similar to sprouts. It is growing in popularity because it contains a fair amount of omega 3 fatty acids. Look for it at your supermarket produce section, greengrocer, or natural foods store. (See recipe for purslane on page 254.) Try to eat ½ cup of this terrific green two to three times a week.

Salmon. Wild Alaskan sockeye salmon is a terrific source of omega 3 fatty acids, especially DHA, and is low in mercury and other toxins.

Study after study confirms that people who consume the most omega 3 fatty acids—and who eat the most fish—are the happiest, healthiest, smartest people on the planet. Try to eat salmon (one serving is 6 ounces) up to twice a week.

Spinach. Popeye ate spinach to improve his physical prowess, but spinach may also give you more mental muscle. Rats fed a diet supplemented with spinach learned to associate the sound of a tone with an oncoming puff of air faster than those fed a nonsupplemented diet. In other words, spinach made rats smarter. Spinach contains lutein, another powerful antioxidant, and is also rich in folic acid, a B vitamin that helps control homocysteine. (As more and more studies are done on specific antioxidant-rich foods, it's possible that other fruits and vegetables will be shown to have specific brain-boosting properties.) Spinach can taste great if it's prepared well. Organic baby spinach tastes the best. Eat it raw in a salad, or sauté it lightly in olive oil over a low flame with garlic slices until it is wilted. Frozen spinach is fine as long as you use an organic brand. Try to eat two to three servings (serving is 1 cup) of spinach a week.

Seeds. Pumpkin seeds and ground flaxseed contain high amounts of omega 3 fatty acids and beneficial GLA. Eat an ounce (a small handful) of pumpkin seeds as a snack; sprinkle them on your salad for flavor and texture. Ground flaxseed (1 ounce) can be added to your protein shake or sprinkled on cereal or a salad. Try to eat two to three servings of pumpkin seeds a week.

Walnuts and walnut oil. These nuts contain high amounts of omega 3 fatty acids. Eat them as a snack, or use walnut oil on your salad. If you are used to snacking on bad stuff, try snacking on a few walnuts and a fresh organic apple. It's a delicious, healthy alternative to junk food. Try to eat two to three servings of walnuts or walnut oil a week. (1 ounce is one serving for nuts; 2 tablespoons is one serving of oil.)

Are You Gluten Intolerant?

Are you suffering from memory loss, confusion, or other neurological symptoms? Do you have chronic upset stomach, or have you experienced unexplained weight loss or bone pain? Any of these symptoms could be a sign of gluten intolerance, a condition that affects about 1 in 250 Americans and is also called celiac disease. Gluten is a protein found in many different grains, including wheat, rye, barley, buckwheat, amaranth, and spelt. When people who are gluten intolerant eat foods containing gluten, their immune cells target this protein as they would a bacteria or virus. This creates an inflammatory response that eventually destroys the villi, the tiny, fingerlike protrusions in the lining of the intestine through which all our nutrients are absorbed. This inflammatory response can directly affect brain tissue much like what is seen in MS, including telltale white lesions in the brain that are similar to the kind of damage seen in MS patients. Gluten intolerance is effectively treated by eliminating gluten from your diet.

Are you gluten intolerant? You can find out by having your doctor give you a simple blood test called the antigliadin antibody test, which measures antigliadin antibodies in the blood, a sign of gluten sensitivity. If you are experiencing any neurological symptoms that suggest MS, I urgently recommend the gluten test. (See chapter 16 for more information.)

What's a Portion?

Tip

1 cup of cereal, chopped vegetables, grains, chopped fruit, etc., is about the size of a closed fist.

6–8 ounces of meat, fish, poultry, or tofu is about the size of the palm of your hand.

½ cup of cottage cheese or yogurt is about the size of a tennis ball.

For the types of food that follow, the amounts listed correspond to one portion.

Salad: 2–3 cups; *a big salad counts as 2 servings of vegetables*
Vegetables: 1 cup raw, ½ cup cooked
Fruit: one medium-sized fruit (apple, pear, or peach), about the size of a tennis ball; 1 cup raw fruit, sliced; ½ cup cooked fruit
Meat, fish, poultry, or tofu: 6–8 ounces (about the size of the palm of your hand)
Protein Powder: 2 tablespoons
Bread and grains: 1 cup of cooked cereal; 1 cup of rice or pasta; 1 medium-sized potato
Eggs and dairy: two large eggs; 1 cup of yogurt, cottage cheese; one slice of cheese, or one piece the size of one-third of your fist
Seeds and Nuts: small handful, or about 1 ounce; 1 tablespoon of nut butter

Foods to Build a Better Brain

Make your daily food selections from the lists that follow. They provide the raw ingredients to build a better brain so that you can achieve your optimal mental performance and maintain it for years to come. The suggested menus that follow show how easy it is to put together delicious meals from these brain-healthy foods. (I also list "brain busters," foods that you should eat rarely, if at all.)

Once you get the bad foods off your plate (the bad fats, the highly processed starchy and sugary foods, and chemical-laden foods) you can pretty much eat what you want, within limits.

Watch your portion sizes! Obesity is probably the number 1 health threat to your brain, and you can avoid it by eating the right amounts of the nutrient-rich foods that I recommend.

Your brain needs a steady supply of fuel throughout the day; therefore, I recommend that you eat at least three meals and a snack or two to keep your brain energized and alert. I've included a list of healthy snacks.

Start your day with some protein, such as eggs, fish, tofu, or a protein shake. Your brain will work better, and you won't be as hungry later in the morning. If you love cereal, choose a high-fiber variety (such as buckwheat or steel-cut oatmeal) but add some protein in the form of nuts or a scoop of protein powder. If you add some flaxseed or flaxseed oil, you also get some good fat.

Ideally, you should have two to three servings of protein a day.

Try to have at least two to three servings of fruit and five to six servings of vegetables daily. Remember that at every meal there should be many more vegetables on your plate than meat.

Fish and Seafood

One serving is 6–8 ounces or about the size of the palm of your hand. Eat two to three fish meals a week.

Fish with the highest omega 6 content are designated by an asterisk. Don't eat farm-raised fish unless it is certified organic. Avoid fish with the highest mercury levels (see chapter 8).

Bass
Cod
Crab
Flounder
Grouper
Haddock
Herring*
Lobster
Mahimahi
Orange roughy*
Oysters (canned or fresh)
Perch
Pike

Pollack
Rainbow trout
Alaskan sockeye salmon
    (canned or fresh)*
Sardines (canned in olive oil)**
Scallops
Shrimps (canned or fresh)
Snapper
Sole
Tilapia
Tuna (light, not albacore,
    canned or fresh, in oil)*
Turbot

*Fish with highest omega 6 content

Meat

One serving is generally considered an amount about the size of the palm of your hand.

Even the cuts of meat that are lowest in fat have a fair amount of saturated fat; therefore, try not to eat more than *two meat meals a week.* Stick to free-range, hormone-free organic meat if you can.

BEEF
Beef tenderloin
Cubed steak
Filet mignon
Flank steak
Ground round, extra lean
Ground sirloin, lean
Round steak
Roast beef (top round
    or rump)
Sirloin steak

LAMB
Chop
Leg
Roast

PORK
Lean, boiled ham
Loin chop
Pork tenderloin

Poultry

One serving is 6 ounces, or roughly the size of two decks of cards.

Stick to organic, free-range, hormone-free poultry. Poultry is low in saturated fat. *You can eat poultry every day if you wish.*

Chicken breast, no skin
Chicken sausage
Ground chicken
Turkey breast

Turkey bacon
Turkey sausage
Ground turkey

Game Meats

One serving is 6–8 ounces, or roughly the size of the palm of your hand.

Stick to organic, free-range, hormone-free game. Game meats are very low in fat, and contain some omega 3 fatty acids. *You can eat game meats every day if you like.*

Cornish game hen, no skin   Pheasant
Buffalo   Rabbit
Ostrich   Venison

Meat Alternatives

One serving of tofu is the size of a deck of cards. One serving of soy or veggie burger is 1 burger.

    Tofu and soy foods are filled with protective antioxidants for your brain but also contain plant estrogens that may increase the risk of certain estrogen-sensitive cancers. *Limit your intake of soy foods to two servings weekly.*

Tempeh (fermented soy)   Boca Burger
Tofu (plain or herb flavored)   Natural Touch garden burger

Seeds, Nuts, and Nut Butters

One serving is 1 ounce or one small handful. One serving of nut butter is 1 tablespoon.

    Stick to unroasted and unsalted seeds and nuts. *You can eat 1 serving of nuts and 1 serving of seeds daily.*

Almonds	Pistachios
Brazil nuts	Pumpkin seeds
Cashews	Poppy seeds
Hazelnuts	Sesame paste
Macadamia	Sesame seeds
Peanuts	Sesame tahini
Pecans	Walnuts
Pine nuts	

Oils

One serving is 1 tablespoon.

Eat two to three servings of good oil daily, in addition to your daily Arctic cod liver oil supplement. (1 tablespoon daily)

Butter (use clarified butter or ghee)

Canola oil

Olive oil

Flaxseed oil

Hemp oil

Mayonnaise (homemade only! See recipe on page 261.)

Pumpkin seed oil

Walnut oil

Eggs and Dairy

One serving is two large eggs; one cup of yogurt or cottage cheese; one slice of cheese; one chunk about the size of half a deck of cards.

You can eat two to three servings of eggs daily. You can eat one to two servings of dairy daily. Full-fat cheese can be high in saturated fat. Stick to no-fat or low-fat varieties with no more than 5 grams of fat per slice or serving.

Eggs (omega 3–enriched eggs only!)

Raw milk goat cheese

No-fat or low-fat (1 percent) cottage cheese

No-fat or low-fat (1 percent) cream cheese

No-fat or low-fat (1 percent) ricotta cheese

No-fat or low-fat (1 percent) yogurt plain, not flavored

Yogurt cheese (lower in fat than milk cheese)

Low-fat cheddar

Low-fat Colby

Low-fat Havarti

Low-fat Monterey Jack

Low-fat provolone

Low-fat Swiss

Protein Powder

One serving is 2 tablespoons in your protein shake or hot cereal.

You can drink one protein shake daily as a snack or meal substitute. Do not eat protein bars. Most brands contain high amounts of sugar or are loaded with chemical additives.

Whey protein

Vegetables

One serving is 1 cup raw (about the size of a closed fist) or $\frac{1}{2}$ cup cooked; 2 cups of salad (no limits on salad!); one medium portion of corn, potato, or yam.

Limit your servings of starchy vegetables (corn, potatoes, yams, turnips) to one serving daily. *Buy organic produce whenever possible.* Peel the skin off produce that is not organic.

Asparagus
Artichoke hearts
Arugula
Avocado
Bamboo shoots
Beets
Bell peppers (red, green, yellow, orange, hot)
Bok choy
Broccoli
Brussels sprouts
Cabbage
Carrots
Cauliflower
Celery
Chard
Chives
Cilantro
Corn
Cucumbers
Endive
Eggplant
Fennel
Greens (collard, turnip, mustard, chard)
Hot peppers
Kale
Kohlrabi
Lettuce (all varieties)
Leeks
Mushrooms (Portobello, shiitake, oyster, button)

Okra
Olives (green and black)
Onions
Parsley
Purslane
Radicchio
Radishes
Scallions
Seaweed (dulse, nori,
    hijiki, kombu)
Snowpeas

Spinach
Sprouts (all varieties)
String beans
Summer squash
Sweet potatoes
Turnips
Water chestnuts
Watercress
Yams
Zucchini

Fruit

One serving is 1 medium-sized fruit (apple, banana, pear, orange, etc.); $\frac{1}{2}$ cup of fruit (grapes, cherries, berries); $\frac{1}{4}$ medium-sized melon (cantaloupe, honeydew); 1 cup of diced watermelon.

Although fruit is packed with beneficial antioxidants, it is also high in sugar, which will raise your blood sugar levels. Chronically high blood sugar levels are associated with memory problems. Don't gorge on fruit, but two servings daily of any fruit is fine. *Dried fruit is very high in sugar. Don't eat it.*

Stick to organic fruit whenever possible. Peel the skin off every fruit that isn't organic. Wash organic fruit carefully in warm water.

Apples
Apricots
Bananas
Blackberries
Blueberries
Cantaloupe
Cherries
Grapefruit
Grapes
Honeydew

Kiwis
Lemons
Limes
Nectarines
Oranges
Papayas
Peaches
Pears
Pineapple
Plums

Raspberries
Strawberries

Tomatoes
Watermelon

Grains and Breads

One serving is one slice of bread, one waffle, or three pancakes (4 inches in diameter and ¼ inch thick); 1 cup cooked cereal, pasta, or rice.

Whole grains are a terrific source of B vitamins, which help to keep homocysteine under control. Grains also contain fiber, which helps slow down the absorption of food and maintain normal blood sugar levels. I recommend that you try breads made with sprouted grains. They are broken down by the body at a slower pace than bread made with flour and therefore cause a more modest rise in blood sugar. They are also easier to digest. They are sold in many supermarkets and health food stores. Eat one to two servings of whole grains daily.

Amaranth
Barley
Basmati brown rice
Buckwheat
Whole wheat couscous
Millet
Oatmeal (steel cut,
   not instant)

Quinoa
Spelt pasta
Sprouted bread or bagels
Vons omega 3–enriched
   waffles
Whole-grain sprouted bread

Legumes

One serving is ½ cup of cooked beans.

Legumes are packed with folic acid, an important B vitamin that helps to control homocysteine levels. They are also low in saturated fat, and an excellent alternative to meat. Combine a serving of legumes with whole wheat pasta or brown rice for a balanced meal. You can eat one to two servings of legumes daily. Vegetarians who rely on legumes for protein can eat three to four servings daily.

Adzukis   Lima beans
Black beans   Navy beans
Garbanzos (chickpeas)   Mung beans
Lentils   Pinto beans

Beverages

Beer (No more than two 8-ounce glasses daily)
Club soda or seltzer (unlimited, counts as water)
Coffee (up to two 6-ounce cups daily)
Tea, green or black (two to three cups daily)
Herbal tea (unlimited)
Spirits (whiskey, gin, vodka): no more than one shot glass daily.
   Avoid mixed drinks made with sweetened mixes. They are loaded
   with sugar.)
Water (bottled or filtered, at least eight 8-ounce glasses daily.)
Wine (red wine is best): no more than two 3-ounce glasses daily)

Condiments, Spices, and Seasonings

Dried and fresh herbs and spices (turmeric, curry, cinnamon, basil, dill, pepper, fennel, ginger, oregano, paprika, cumin, etc.) contain high amounts of antioxidants, which is why they were often used as preservatives in the days before refrigeration. The same free radicals that turn food rancid can turn your brain cells rancid. Use fresh herbs and spices liberally in your cooking.

All dried herbs and spices
Bragg's Liquid Aminos (nonfermented soy sauce substitute)
Capers
Celtic salt (No more than 1 tablespoon of added salt daily to your
   food. Avoid all salt if you are on a salt-restricted diet).
Garlic (fresh or powdered)
Miso (only if you are not on a salt-restricted diet)
Mustard

Reduced sodium soy sauce or tamari

Vinegar (balsamic, red wine, umeboshi plum, and rice)

Great Snacks to Feed Your Brain

One small handful of walnuts

One small handful of pumpkin seeds

One fresh pear with a portion of nuts or seeds

Fruit smoothie with ground flaxseed

Guacamole with walnuts

Garlic hummus with celery

Fruit and walnut chews

Walnut flax bar

See the snacks recipes in Appendix 2.

Brain Busters: Foods You Should Avoid

Fish with the Highest Mercury Content

These fish contain the highest mercury levels. I don't recommend that you eat them. *Pregnant women and children should avoid these fish.*

Halibut	Swordfish
King mackerel	Tilefish
Shark	Tuna (white meat)

Fatty Cuts of Meat

Avoid these meats; they contain a huge amount of saturated fat, which will prevent the good fat from getting into your brain cells. *Eat them rarely (not more than two to three times a year) if at all.*

Bacon (pork)	Knockwurst
Bratwurst	Ground pork
Brisket	Pork sausages
Deli meats (pastrami, corned	Spareribs
beef, salami, bologna)	Prime rib
Full-fat pork hotdogs	Rib steak

Say No to Fried Foods

Fried foods are high in trans-fatty acids, which make for terrible brain cells and sluggish thinking. Don't eat them!

Any fried or deep-fried food	Chicken nuggets
Fried chicken	French fries
Fried fish	

Refined Grains and Starches

Anything made with white, processed flour is bad! It raises your blood sugar level (which makes you more susceptible to memory problems) and is devoid of beneficial brain-building B vitamins.

White bread
White rolls
Corn muffins
Crackers made with hydrogenated oil (nearly all are!)
Sweetened processed cereal
Frozen waffles made with hydrogenated oil

Sweets

A high-sugar diet can accelerate memory loss and destroy delicate nerve cells. Try not to eat more than one high-sugar dessert a week. Try to avoid the following foods.

Cupcakes
Dried fruit
Frozen custard
Gelato
Frozen yogurt

Fruit-flavored yogurt
Fruit rolls
Marshmallows
Ice cream
Sorbet (very high in sugar)

Bad Fats

Commercially processed oils contain trans-fatty acids and/or are highly vulnerable to oxidation. I don't recommend that you eat or use in cooking any of the fats listed here.

Corn oil
Hydrogenated or partially hydrogenated fats
Lard
Margarines with trans-fatty acids
Peanut oil
Soy oil
Squeezable butter or shortening

High-Sugar Condiments

Use sparingly (no more than once a week)

Barbecue sauce (2 tablespoons)
Most commercially prepared salad dressings
Ketchup (2 tablespoons)
Mayonnaise (See the recipe for homemade mayonnaise, page 261.)

Beverages

Nothing with added sugar or artificial sweeteners. Don't drink these beverages.

Fruit juice, all varieties, even fresh squeezed. Eat fresh fruit instead; it contains sugar, but it also contains a high amount of fiber, which offers substantial health benefits.

Soda pop, all varieties, diet or sweetened

Sports drinks

Sweetened teas

Excitotoxins

Aspartame

Hydrolyzed vegetable protein (found in processed foods, frozen TV dinners, canned soups, etc.)

MSG

Sugar and Artificial Sweeteners

Avoid foods containing the following ingredients. If sugar is listed as one of the top three ingredients of a product, it is a high-sugar food and should be eaten sparingly. Using a tablespoon or two of maple syrup on your whole-grain pancakes once or twice a week is acceptable, but eating a highly processed, sugar-rich diet loaded with packaged snack foods, soda, cakes, cookies, chips, and white bread is not.

Artificial sweeteners can overexcite brain cells, causing neurological problems such as confusion and dizziness in some people. I recommend that you don't use them.

People often ask me if any form of sugar is better than another. Any form of sugar raises blood sugar levels, which can harm brain cells and, in particular, cause memory problems. In this regard, so-called natural sugars such as brown sugar or honey are no better than table sugar. Therefore, you must tightly limit your intake of these sugars as well.

Aspartame	Fructose
Corn syrup	Honey
Dextrose	Maple sugar

Maple syrup
Saccharin
Sucrilose

Sucrose
Sugar

SNACK FOODS

Snack foods are loaded with sugar and bad fats, especially trans-fatty acids. You can live without eating these foods.

Chips (corn, potatoes, cheese, etc.)
Breakfast bars
Energy bars
Granola bars
Commercially prepared cakes
Nonorganic, white, processed cake or pancake mix
Candy
Packaged cookies (beware of hydrogenated oils)
Flavored gelatin desserts (sugar free or regular)
Popcorn (many brands are very high in trans-fatty acids)
Pretzels

Meal Plan

The menus that follow were designed by J. Gabrielle Rabner, M.S., R.D., who is based in Naples, Florida, and Montclair, New Jersey. Foods with accompanying recipes are designated by an asterisk. Recipes are in Appendix 2. Portion sizes have been discussed earlier in this chapter.

Week 1

SUNDAY	
Breakfast	• One medium-sized apple, sliced (peel fruit if it is not organic)
	• Two poached omega 3–enriched eggs

	•	Sprouted-grain toast and organic all-fruit unsweetened blueberry jam
	•	Coffee or tea
Lunch	•	Lentil soup with grated raw milk cheese and pumpkin seeds*
	•	Spinach salad with lemon and 1 tablespoon of olive oil
Snack	•	Honeydew melon balls and fresh blueberries
	•	Walnuts
Dinner	•	Chicken with honey mustard sauce*
	•	Baked sweet potato with cinnamon
	•	Mixed green salad with walnut oil and lemon

MONDAY

Breakfast	•	One fresh orange cut into sections
	•	Sprouted-grain bread with walnut butter
	•	Coffee or tea
Lunch	•	Endive and radicchio salad with sliced chicken breast and avocado slices
	•	Mustard vinaigrette
Snack	•	Fresh pear with goat milk cheese and walnuts (peel the pear if it is not organic)
Dinner	•	Haddock stew*
	•	Green salad with grilled red beets
	•	Flaxseed and olive oil dressing*

TUESDAY

Breakfast	•	Protein shake with banana, strawberries, and ground flaxseeds*
	•	Coffee or tea
Lunch	•	Baked garlic and soy sauce tofu*
	•	Raw sliced fennel drizzled with olive oil and lemon
Snack	•	Plain yogurt with pumpkin seeds and apple slices
Dinner	•	Gingered flank steak*
	•	Steamed purslane*
	•	Basmati brown rice

WEDNESDAY

Breakfast
- Fresh pineapple slices
- Hot oatmeal (regular, not instant) with flaxseeds and sunflower seeds
- Coffee or tea

Lunch
- Vegetarian Caesar salad*
- Sliced cold flank steak (leftovers)

Snack
- Fresh apple and walnuts (peel the apple if it is not organic)

Dinner
- Stir-fried vegetables with marinated tempeh*
- Leaf lettuce salad with olive and pumpkin seed dressing*

THURSDAY

Breakfast
- Fresh peach (peel the peach if it is not organic)
- Millet cereal drizzled with flaxseed oil and topped with sliced almonds*
- Coffee or tea

Lunch
- Grilled lean chopped sirloin hamburger
- Oven-roasted vegetables*
- Raw baby carrots

Snack
- Garlic hummus (chickpea dip) with cut-up vegetables*

Dinner
- Roast turkey with shiitake mushrooms
- Grilled turnips
- Steamed kale drizzled with walnut oil

FRIDAY

Breakfast
- Blueberries
- Omega 3–enriched egg omelet with spinach and mushrooms
- Whole-grain toast
- Coffee or tea

Lunch
- Turkey burrito with corn tortilla
- Spinach salad with flaxseed and olive oil dressing

Snack
- Fruit smoothie with ground flaxseeds*

Dinner
- Meatballs over spelt pasta
- Arugula salad with pumpkin seed and olive oil dressing

SATURDAY	
Breakfast	• Buckwheat pancakes with maple syrup and fresh berries*
	• Coffee or tea
Lunch	• Poached wild salmon on a bed of steamed kale
	• Grilled shiitake mushrooms
Snack	• Guacamole dip with walnuts*
	• Celery and carrot sticks
Dinner	• Baked chicken with turnips and parsnips*
	• Steamed collard greens drizzled with walnut oil

Week 2

SUNDAY	
Breakfast	• Tangerine
	• Spinach, mushroom frittata*
	• Coffee or tea
Lunch	• Grilled turkey burger with lettuce and tomato on a spelt roll
	• String beans with pumpkin seeds and garlic vinaigrette*
Snack	• Baked apple with walnuts and figs*
Dinner	• Meatball chili*
	• Green salad with flaxseed oil and olive oil dressing*

MONDAY	
Breakfast	• Fresh pear (peel the pear if it is not organic)
	• Scrambled tofu with onions and broccoli*
	• Sprouted grain toast and butter
	• Coffee or tea
Lunch	• Waldorf salad with homemade mayonnaise*
Snack	• Mixed berries on plain yogurt
Dinner	• Roast Cornish hens
	• Wild rice with pumpkin seeds
	• Endive salad with lemon dressing

TUESDAY	
Breakfast	• Brown rice pudding with apple slices, raisins, and pumpkin seeds*
	• Coffee or tea
Lunch	• Greek salad with olive oil and walnut oil dressing*
Snack	• Fresh pear with goat cheese and walnuts
Dinner	• Grilled filet mignon
	• Sautéed kale in olive oil with garlic
	• Boston lettuce with pumpkin seed oil

WEDNESDAY	
Breakfast	• Rice protein shake with pineapple, apricots, and ground flaxseeds*
	• Coffee or tea
Lunch	• French onion soup*
	• Mixed vegetable salad with walnut oil and balsamic vinegar
Snack	• Cantaloupe and tamari pumpkin seeds*
Dinner	• Stir-fried chicken with vegetables
	• Basmati brown rice

THURSDAY	
Breakfast	• Fresh plum (peel the plum if it is not organic)
	• Soft-boiled omega 3–enriched eggs
	• Sprouted-grain bread with raspberry all-fruit unsweetened jam
	• Coffee or tea
Lunch	• Salade Niçoise on a bed of greens*
Snack	• Black bean dip with vegetable slices*
Dinner	• Baked tilapia*
	• Roasted vegetables and stuffed mushrooms*
	• Whole-grain couscous

Breakfast	• Fresh berries
	• Turkey bacon on spelt bagel with apple butter
	• Coffee or tea
Lunch	• Butternut squash soup garnished with pumpkin seeds*
	• Celery sticks stuffed with raw herb goat cheese
Snack	• Walnut flax bar*
Dinner	• Grilled loin lamb chops with rosemary and garlic
	• Greens with flaxseed oil and olive oil dressing
	• Barley pilaf*

SATURDAY

Breakfast	• Apple walnut soufflé*
	• Coffee or tea
Lunch	• Cold lamb (leftovers) on bed of green with walnut oil and raspberry vinegar
Snack	• Cantaloupe pieces and walnuts
Dinner	• Tarragon chicken with lemon*
	• Baked sweet potato slices
	• Greens with flaxseed oil and olive oil dressing

Week 3

SUNDAY

Breakfast	• Grapefruit slices
	• Deviled omega 3–enriched eggs
	• Tomato slices drizzled with flaxseed oil
	• Coffee and tea
Lunch	• Buffalo burger on sprouted grain roll
	• Red and green cole slaw
Snack	• Walnuts and dried apricots
Dinner	• Baked trout*
	• Butternut squash stuffed with walnuts and apples
	• Mixed green salad with pumpkin seeds and olive oil and lemon

MONDAY	
Breakfast	• Protein shake with berries and flaxseeds
	• Coffee or tea
Lunch	• Turkey breast on mixed green salad
	• Lemon, parsley, and flaxseed oil dressing*
Snack	• Poached pear with walnuts*
Dinner	• Meat loaf
	• Roasted root vegetables*
	• Green salad with poppy seed dressing*

TUESDAY	
Breakfast	• Kiwi
	• Tomato basil omelet*
	• Sprouted-grain toast with grape all-fruit unsweetened jam
	• Coffee or tea
Lunch	• Split pea dill soup with pumpkin seeds*
	• Mixed green salad topped with sheep milk cheese
Snack	• Cranberry flaxseed muffin*
	• Orange tea
Dinner	• Honey-glazed baked wild Alaskan salmon*
	• Wild rice with slivered almonds
	• Steamed Brussels sprouts

WEDNESDAY	
Breakfast	• Apple and pear slices (peel fruit if it is not organic)
	• Buckwheat and flaxseed pancakes with maple syrup*
	• Coffee or tea
Lunch	• Salmon (leftovers) salad on a bed of dark greens
	• Lemon vinaigrette
Snack	• Avocado salsa*
	• Jicama, celery and carrot sticks
Dinner	• Broiled flank steak with tamari soy sauce and garlic marinade*
	• Brown rice with broccoli and walnuts
	• Endive salad and lemon, parsley dressing

THURSDAY	
Breakfast	• Fresh peach (peel fruit if it is not organic)
	• Poached omega 3–enriched egg on millet*
	• Coffee or tea
Lunch	• Quinoa and beef stuffed red pepper*
	• Mixed salad with walnut oil and olive oil dressing
Snack	• Protein shake with banana, papaya, and ground flaxseeds*
Dinner	• Broiled venison cubes or lean beef cubes*
	• Spinach and mushroom salad with pumpkin seed oil dressing

FRIDAY	
Breakfast	• Fresh blueberries
	• Sprouted-grain bread French toast with all-fruit unsweetened jam*
	• Coffee or tea
Lunch	• Black bean soup with pumpkin seeds garnish*
	• Green salad with vinaigrette
Snack	• Fruit and walnut chews (uncooked)*
	• Peppermint tea
Dinner	• Leg of lamb with garlic and rosemary
	• Wild rice and walnut salad
	• Steamed collard greens

SATURDAY	
Breakfast	• Broccoli frittata*
	• Mixed fruit kabobs
	• Coffee or tea
Lunch	• Couscous lamb (leftovers) salad
	• Steamed purslane
Snack	• Celery stuffed with natural peanut butter and topped with flaxseeds
Dinner	• Chicken and ginger sauté over spelt pasta*
	• Green salad with walnut oil and olive oil dressing

Week 4

Breakfast	• Oatmeal (not instant) with flaxseeds and pumpkin seeds
	• Coffee or tea
Lunch	• Lentil soup
	• Pumpkin seed butter on rice crackers*
Snack	• Slice of walnut quick bread*
	• Ginger tea
Dinner	• Salmon cakes*
	• Green salad
	• Roasted butternut squash slices

MONDAY

Breakfast	• Protein shake with peach, grapes, and ground flaxseed
Lunch	• Tempeh burgers with lettuce and tomato on sprouted grain roll
	• Celery and carrot sticks
Snack	• Gazpacho soup with walnut garnish (Gazpacho soup can be purchased in most gourmet shops and health food stores and in the prepared food section of most supermarkets. Serve cold, floating a few walnut pieces in the soup.)
Dinner	• Roast turkey breast with turnip slices
	• Whole-grain couscous
	• Sautéed green beans with olive oil and walnut oil dressing

TUESDAY

Breakfast	• Apple and raw milk goat cheese (peel apple if it is not organic)
	• Flaxseed crackers
	• Coffee or tea
Lunch	• Turkey (leftovers) on a bed of greens with red onions and sliced tomatoes
	• Flaxseed oil and olive oil dressing*
Snack	• Fresh carrot and spinach juice shake with ground flaxseeds and rice protein*

Dinner	• Chicken stir-fry with walnuts and vegetables*
	• Brown rice
	• Arugula and watercress salad with olive oil and flaxseed oil dressing

WEDNESDAY

Breakfast	• Protein shake with mango, cherries, and ground flaxseeds
Lunch	• Buffalo beef chili (substitute buffalo meat in your favorite chili recipe)
	• Mixed green salad with olive oil and flaxseed oil dressing
Snack	• Eggplant tahini dip with pumpkin seeds and assorted vegetables
Dinner	• Broiled tofu with walnut sauce
	• Brown rice with collard greens

THURSDAY

Breakfast	• Omega 3–enriched eggs stuffed with spinach and flaxseeds
	• Coffee or tea
Lunch	• Tilapia fish chowder*
	• Spinach salad with lemon and miso dressing*
Snack	• Whey protein shake with pineapple, strawberry, and ground flaxseeds*
Dinner	• Lamb curry with walnuts*
	• Cucumber salad with pumpkin seeds

FRIDAY

Breakfast	• Baked eggs Françoise*
	• Whole grain toast and blackberry all-fruit unsweetened jam
	• Coffee or tea
Lunch	• Pumpkin soup garnished with seeds*
	• Endive and radicchio salad with olive oil and walnut oil dressing*
Snack	• Apple and walnut squares*
Dinner	• Crispy oven-baked chicken with chunks of sweet potatoes
	• Leaf and frisee lettuces with lemon and walnut vinaigrette*

SATURDAY		
Breakfast	•	Fresh figs
	•	Millet pancakes*
	•	Coffee or tea
Lunch	•	Chef's salad with chicken (leftovers) and pumpkin seed vinaigrette*
Snack	•	Walnut pâté* with fresh pear slices
Dinner	•	Venison steak with mushrooms and onions
	•	Roasted beets
	•	Greens with walnuts and lemon vinaigrette

Tailor a Supplement Program to Your Risk Factors

SUPPLEMENTS are vitamins, minerals, amino acids (the building blocks of protein), herbs, essential fatty acids, and phytochemicals (extracts from plants) that work in synergy with food to recharge your brain and protect it against free radical damage and inflammation. Supplements cannot replace a good diet (see chapter 5), but they can compensate for nutrient deficiencies that are a result of a poor diet or drugs that deplete the body of vital nutrients.

The Better Brain supplement regimen is organized in three tiers based on risk factors. Your score on the Brain Audit will determine the right tier for you. Tier 1, Tier 2, and Tier 3 address specific risk factors with specific supplement regimens (provided at the end of this chapter and in Appendix 1). These are the same regimens that I use for my patients and that yield consistently excellent results. If you or a loved one have been diagnosed with Alzheimer's disease, Parkinson's disease, stroke, vascular dementia, ALS, or multiple sclerosis, you should read Part 3. Although you will follow the Tier 3 program, there may be additional supplements or different dosages depending on your particular problem.

The supplements described here are the same supplements I use in my practice. In the pages that follow, I explain why I consider them to be so essential for brain health, and how to use them safely and effectively.

Acetyl-L-carnitine
Alpha lipoic acid
Vitamin B complex
Vitamin C
Coenzyme Q10 (Co-Q10)
Vitamin D
DHA
Vitamin E
Folic Acid
Ginkgo Biloba
N-acetyl-cysteine (NAC)
Phosphatidylserine (PS)
Vinpocetine

Acetyl-L-carnitine

DAILY DOSE

Tier 2: *400 mg once daily.*
Tier 3: *400 mg twice daily for a total of 800 mg*

If you're stressed out or lacking in mental energy, this supplement is just what the doctor ordered. Carnitine is an amino acid, a building block of protein that is naturally produced by the body and found in food. Acetyl-L-carnitine is a chemically active form of this amino acid that is sold as a supplement. Carnitine is very safe—it's even present in breast milk and is added to infant formula because it is so essential for normal physical and mental development. Adults need carnitine as much as infants do, but carnitine levels decline with age.

Carnitine is important for optimal brain function for several reasons. First, carnitine is readily converted into acetylcholine, a neurotransmitter that is critical for learning and concentration. If you are lacking in acetylcholine, you will not be able to perform at a minimal mental level, let alone at your peak. Second, carnitine is a *neuronal energizer;* it is essential for the production of energy by the mitochondria, the "power-house" of the cell. Carnitine transports fatty acids across the mitochondrial membrane, where they are made into energy. Carnitine also helps remove waste products from mitochondria-created energy production, enabling them to be eliminated from the body. This is a very important job; if toxins are not removed from mitochondria, they can damage the mitochondria, which will slow down energy production even more. Without enough energy, your brain cells can't produce enough neurotransmitters (like acetylcholine) or communicate quickly or effectively with each other. Without enough energy, your brain can't produce enough antioxidants to protect it against free radical damage, and your mental performance will inevitably suffer. As more and more brain cells are destroyed by free radicals, you will have more difficulty concentrating, your response time will slow down, and you will find yourself wasting precious time hunting down your glasses and your car keys.

If you are low in carnitine, you will also be more vulnerable to the toxic effects of stress. Chronic exposure to cortisol, a hormone produced by the adrenal glands during times of stress, can injure brain cells in the hippocampus, the memory center of the brain; this is a common cause of short-term memory loss. When we are young, we are protected against excess cortisol production. When the adrenal glands go into overdrive, special cortisol receptors in the brain alert them that it's time to shut off cortisol production. As we get older, that is, as free radicals begin to damage healthy cells, these receptors become dulled and do not respond as quickly or as well. Therefore, our brains are exposed to higher levels of cortisol for extended periods of time. Supplementation with acetyl-L-carnitine can help restore cortisol receptors in the brain that help protect us against the toxic effects of stress.

Carnitine is one of the few substances that can help slow down the progression of Alzheimer's disease. People with Alzheimer's disease have strikingly low levels of carnitine. In one recent study, researchers at the University of California at San Diego found a dramatic reduction in the rate of mental decline among younger patients with Alzheimer's who were taking acetyl-L-carnitine supplements over the one-year evaluation period. Other studies have found that acetyl-L-carnitine can help improve mood among Alzheimer's patients. Many researchers believe that the age-related decline in carnitine may be a causal factor in the onset of Alzheimer's disease, which suggests that supplementing with this vital amino acid in midlife, when levels begin to decline, may help prevent this brain degenerative disease in the first place.

Alpha Lipoic Acid

DAILY DOSE

Tier 2: *80 mg*
Tier 3: *200 mg*

Are you locked into the "What was your name again. . . ? Where did I put my keys?" syndrome? Consider taking alpha lipoic acid, an antioxidant that is made by the body and is also critical for energy production in the brain.

In the search for the ultimate "smart drug," researchers are constantly testing substances on animals to find one that can enhance memory without damaging delicate brain cells. Most of these tests fail miserably—a drug that is strong enough to alter brain chemistry is often toxic to brain cells—but there is solid evidence that alpha lipoic acid in combination with carnitine (see the preceding section) can do the job safely and effectively. A team of researchers at the University of California at Berkeley, led by Bruce N. Ames, fed older rats two dietary supplements, alpha lipoic acid and acetyl-L-carnitine. Animals taking both supplements not only performed better on memory tests as compared to animals given a placebo, but they appeared to be friskier and more energetic than the untreated animals. Analysis of brain tissue showed less damage to the mitochondria, the energy-producing center of the cell in treated animals, and less oxidative damage to the hippocampus, the memory center of the brain. These two supplements had a rejuvenating effect on the older animals by stopping free radical damage and revving up energy production. Since the same mechanisms that age rat brains can age human brains, it's not a great leap of faith to conclude that these two supplements can protect human brains from the ravages of age.

Alpha lipoic acid is one of the few antioxidants that can significantly boost glutathione, the brain's most important antioxidant. Glutathione offers powerful protection against the destruction caused by free radical attack on delicate brain cells. If your brain is being devoured by free radicals, you will not be able to think clearly, stay focused, or retrieve information when you need it. The problem is, glutathione levels decline with age, and glutathione supplements are poorly absorbed. Taking alpha lipoic acid is an effective way to maintain glutathione and protect your brain from damage.

If you have been exposed to high levels of toxic metals (such as lead) take note: alpha lipoic acid is an excellent chelating agent, which means which means that it binds to potentially toxic metals including iron and helps eliminate them from the body. (For more information on the effect of toxins on the brain, see chapter 8.)

Vitamin B Complex

DAILY DOSE

Tier 1: *1 B-complex supplement daily*
Tier 2, Tier 3: *One B-complex supplement daily plus additional folic acid and vitamin B12*

A basic B-complex supplement should include the following:

B1 (thiamine)	50 mg
B3 (niacin as niacinamide)	50 mg
B6 (pyridoxine)	50 mg
Folic acid	400 mcg
B12 (cobalamin)	500 mcg

There are several brands of B complex that offer these supplements in one capsule.

If you are irritable, unhappy, and really feel like you're "losing it," chances are you are low on one or more B vitamins. Low levels of B vitamins are typical among people suffering from depression, memory problems, and dementia. B vitamins are critical for brain health primarily because they control homocysteine, the amino acid naturally produced by the body. High levels of homocysteine can promote inflammation, damage blood vessels delivering blood to the brain, and kill brain cells. Researchers have reported a clear link between elevated homocysteine and decreased mental performance, even in seemingly healthy adults. In particular, elevated homocysteine can affect areas requiring psychomotor speed or reaction time in performing specific tasks. Psychomotor speed is particularly important for activities involving coordinated hand-eye movements such as driving a car or playing golf or tennis. If you don't want to lose your psychomotor speed skills, take your B vitamins and have your physician check your homocysteine levels each year.

B vitamins are ubiquitous in the food supply: they are found in meat, fish, eggs, whole grains, fortified cereals, and in lesser amounts in fruits and vegetables. The problem is, B vitamins can be quite fragile and can be destroyed by microwave cooking or high heat. Because they are water soluble, they can be lost in cooking fluid. Thiamine is destroyed by alcohol and therefore is often deficient in alcoholics. Since animal products are the best source of many B vitamins, vegetarians are at particular risk of running low on them, especially B12.

Warning: Dozens of commonly used drugs, from estrogen to antacids, can deplete your body of B vitamins. See chapter 4 for a list of drugs that can sap your body of vitamin B. If you are taking any of these drugs, you may need to add additional B vitamins to your supplement regimen (see chapter 4 for more details).

I recommend that everyone take a basic B-complex supplement. In one pill, you can get much of the B vitamins you need to keep your brain functioning well. Those who are following the Tier 2 and Tier 3 programs will need to take additional doses of folic acid and vitamin B12 (see the following).

Vitamin B12 (Cobalamin)

Tier 2, Tier 3: *You may need to take an additional 500 mcg of vitamin B12 along with your B-complex supplement*

If you pop antacids throughout the day or are a strict vegetarian, you may be deficient in this important B vitamin. Low levels of B12 can result in symptoms such as confusion, depression, and difficulty maintaining your balance.

Why is B12 so important? It is essential for the formation and maintenance of myelin, the protective covering around nerve cells (brain cells are nerve cells). Myelin breaks down during the aging process, which can damage brain cells and slow you down both mentally and physically, depending on which part of the brain is most affected. B12 is also part of the group of B vitamins that helps police homocysteine, the amino acid found in the body, that if produced in excess can damage your brain cells by promoting inflammation in your brain.

Vitamin B12 is found in foods of animal origin: eggs, meat, fish, and dairy products. Vegetarians—especially vegans who don't even eat eggs—are at special risk of being deficient in this vitamin and should take at least 500 micrograms of B12 daily. B12 deficiency is also quite common among people over 60, who are prone to develop a condition called atrophic gastritis, characterized by less gastric acid and reduced production of "good" bacteria that help to break down protein, especially in meat, the major source of B12 in the diet. The symptoms of atrophic gastritis are similar to those of indigestion, and many people mistakenly treat it by taking antacids, which will make even less gastric acid available for digestion and only aggravates the problem. Symptoms of B12 deficiency include confusion, numbness or tingling in the arms or legs, balance problems, and even dementia. Unfortunately, these symptoms may be dismissed by family members or even physicians as a "normal" consequence of aging. They most definitely are not. And when a patient presents with these symptoms, one of the first things I do is check their B12 levels. B12 shots and supplements can work wonders in restoring these patients to their normal selves.

Vitamin C

DAILY DOSE

Tiers 2 and 3: *200 mg twice daily for a total of 400 mg daily*

Vitamin C is well known for its immune-boosting properties, but I would like to give yet another reason to take this powerful antioxidant: it is essential for good brain health. Similar in molecular structure to glucose, the brain's fuel, vitamin C easily crosses the brain-blood barrier. There is a huge concentration of vitamin C in the fluid around the neurons in the brain. This makes perfect sense—vitamin C regenerates, or boosts the effectiveness of, another antioxidant that is important to the brain, vitamin E. If you don't have enough vitamin C in your brain, your brain will be more vulnerable to free radical attack and the downward spiral that leads to diminished brain function and a poor quality of life. The minor memory and cognitive function problems that begin in midlife are a sign that your brain is under free radical attack. Vitamin C is an important tool to stop and reverse some of the damage that has already been done, and to prevent future damage.

There is compelling evidence that simply taking your antioxidant supplements now will protect you from brain aging later. A study involving nearly 3,400 Japanese-American men from the Honolulu Aging Study found that older men who took supplements of both vitamin C and vitamin E at least once a week were protected from vascular dementia, which is caused by impaired blood flow to the brain, and performed better on memory tests than men who did not take vitamins C and E. The men who took vitamins C and E together for many years had significantly better test results than those who took them later in life. *Men who regularly took C and E supplements were 88 percent less likely to develop vascular dementia than those who did not.* Vascular dementia is not some rare disease—it's the leading cause of senility, second only to Alzheimer's disease. If simply taking a few pills a day can drastically reduce your risk of this problem down the road, why not do it?

Coenzyme Q10

DAILY DOSE

Tier 1: *30 mg*
Tier 2: *60 mg*
Tier 3: *200 mg*

Co-Q10 is so critical for brain health that it is one of the few supplements that I recommend to everyone. That's why it is part of the Tier 1 program, which is designed for maximum maintenance and prevention. There is no doubt in my mind that if more people took Co-Q10, starting in young adulthood, there would be fewer older adults with failing brains and nerve cells. It is also part of the Tier 2 and Tier 3 programs, both designed to restore lost function and prevent further damage. There is no doubt in my mind that if more people took Co-Q10 at the first sign of brain decline, they could reinvigorate their flagging mental function and spare their brains from further damage.

Co-Q10 is found in food and is produced by every cell in the body, although there are particularly high concentrations of it in the brain. Production of Co-Q10 declines with age, and there is a growing belief among the scientific community that the loss of Co-Q10 is a prime cause of the decline in mental function in midlife. In the body, Co-Q10 is synthesized from the amino acid tyrosine, with the help of about a dozen other vitamins. If you are low in any of the essential components of Co-Q10, which is possible, given the nutrient-poor modern diet, your body may not make enough of it.

Why do I think Co-Q10 is so important? Let me explain what it does and what makes it so unique. Co-Q10 is involved in the production of energy in the mitochondria, the energy-producing structures within cells. Co-Q10 has been dubbed the cellular spark plug because just as a spark plug is needed to start the engine of an automobile, Co-Q10 is essential for the production of energy in the cell. Since the brain is one of the most metabolically active organs in the body, it needs lots of Co-Q10 to make the energy to perform all its vital tasks. You can't think, learn, or remember things if your brain cells are short of fuel, and you can't have enough fuel if you don't have enough Co-Q10.

Co-Q10 is not just vital for energy production but is also an antioxidant that can protect cells from the negative byproducts of energy production—free radicals. Without enough Co-Q10, your brain cannot produce enough energy, nor will you have enough protection against free radicals. Without enough Co-Q10, your brain will degenerate faster, and you will find it increasingly difficult to maintain your mental performance at an optimal level.

Warning: Dozens of commonly used drugs, from antidepressants to cholesterol-lowering medications, deplete the body of Co-Q10. If you are taking a drug that destroys Co-Q10 you must compensate by taking additional Co Q10 supplements (see chapter 4).

Co-Q10 has recently been touted as one of the few effective treatments for Parkinson's disease, a disease characterized by low energy production and high free radical activity in the brain. (For more information on Parkinson's disease, see chapter 15.) At the Perlmutter Health Center, we've been prescribing Co-Q10 to our Parkinson's patients for more than a decade, with good results. My favorite study involving Co-Q10 was published in the British medical journal *Lancet* about 10 years ago. In the study, patients suffering from a severe, inherited form of dementia were given Co-Q10, along with supplemental vitamin B6 (pyridoxine) and iron. Patients' symptoms abated during the time they were taking the supplements but returned once they discontinued the supplements. Describing one of the patients in the study, the researcher wrote: "Her daily activity improved from Stage 5 (moderate Alzheimer's disease) to 1 (normal). She had increased blood flow to the cerebral cortex and decreased symptoms of clinical dementia. . . . *She now rides a motorcycle.*"

Vitamin D

If you're moody or worried about brain aging or both, be sure to take your vitamin D. You may think of vitamin D only as the companion vitamin you take with calcium to build strong bones, but in reality it is one of the most potent antioxidants on the planet. It is an even more powerful antioxidant than vitamin E in terms of protecting cell membranes from free radical attack, which is the root cause of all mental performance problems, from the minor everyday memory mishaps to severe forms of dementia. Vitamin D can also boost glutathione production in the liver, which will enhance the body's detoxification system. This helps your brain by protecting delicate neurons from toxins in food, drugs, and the environment that promote free radical production and inflammation.

Vitamin D is also essential for a good mood, particularly if you find yourself feeling down during fall and winter. Many people suffer from seasonal affective disorder (SAD) in the fall and winter when there is less exposure to sunlight. Why? Sunlight triggers the production of serotonin, a chemical in the brain that helps regulate mood, among other things. A recent placebo-controlled, double-blind study conducted in Australia found that giving SAD sufferers 400 IU or 800 IU of vitamin D daily for five days in late winter made them feel better. If you have problems with SAD, try taking vitamin D for a few weeks and see if it works for you.

I recommend vitamin D supplements for people following the Tier 2 and 3 program. If you are already taking a multivitamin or a calcium supplement, please check first to see if your supplement includes vitamin D. In very high doses it can be toxic, and I don't recommend that you exceed 400 IU daily.

Docosaheaxaenoic Acid (DHA)

DAILY DOSE

Tier 1 and Tier 2: *300 mg*
Tier 3: *300 mg twice daily*

Your mental performance is dependent on the quality of your brain cells, and the quality of your brain cells is dependent on whether you are feeding them enough DHA, which is an essential fatty acid that is absolutely critical for a well-functioning brain. This supplement is so important that it is one of the few supplements that I recommend for all three tiers of the Better Brain program.

What happens when you don't have enough DHA in your brain? You'll be depressed, you won't be as smart or alert as you can be, and you'll be at greater risk of developing Alzheimer's disease. Not a pretty picture! Chances are that unless you are taking DHA supplements or making a real effort to put DHA-rich foods on your plate, you are not getting enough DHA for optimal brain performance.

Your brain needs DHA and a lot of it. About 25 percent of the total human brain fat is composed of DHA, especially the cell membrane, which is vitally dependent on adequate amounts of DHA for optimal functioning. DHA provides brain cell membranes with the flexibility necessary for efficient communication; it enables brain cells to pass information so that you can think better and faster. If there isn't enough DHA on hand to repair and make new brain cells, your brain will use bad fats like saturated fat and trans-fatty acids, which can make the brain cell membrane hard and rigid. If the cell membrane contains inferior fats due to a shortage of DHA in the diet, the brain's communication network will be compromised. As I tell my patients, "sluggish" fat produces a "sluggish" brain, and good fat produces a great brain.

DHA is not produced by the body, and must be obtained through food or supplements. (DHA is abundant in fatty fish such as salmon, tuna, and sardines.) The problem is, many people do not get enough DHA in their diet, and as a result their brains do not work as well as they should. Over the past one hundred years, due to modern food-processing techniques, our consumption of omega 3 fatty acids has declined significantly in the Western world. As a result, few people can consume enough DHA through diet alone. DHA production may also be hampered by other factors, including excess alcohol consumption and a high intake of saturated fat and trans-fatty acids.

In a groundbreaking article published in the *American Journal of Clinical Nutrition,* Drs. Joseph R. Hibbeln and Norman Salem Jr., of the National Institutes of Health, said that the increase in depression reported in North America during the twentieth century was probably due to the decrease of DHA in the diet. In a recent report in the *Archives of General Psychiatry,* researchers demonstrated a remarkable 50 percent reduction in the standard Hamilton Depression Rating Scale in 54 percent of the research subjects taking a fish oil supplement with DHA, providing scientific proof that DHA can elevate mood.

The dietary deficit in DHA may have a profound effect on future generations. Recent studies show that the amount of DHA in the breast milk of American women is among the lowest in the world. In children, low levels of DHA have been linked to violent behavior, learning disorders, depression, and visual problems. A recent study showed that the IQs of formula-fed infants were eight points lower on average than those of breast-fed infants. Researchers concluded that the difference in IQ was due to the absence of DHA in breast milk. The World Health Organization has recommended that all infant formulas be DHA-enriched, and many manufacturers have already added DHA to their infant formula. DHA is so vital for your brain that I recommend DHA supplements for people of all ages, in all stages of life.

Since DHA is often derived from fish, some brands may contain mercury or other toxins. Since DHA is a fat, it is prone to oxidation, and therefore I recommend specific brands. (See Appendix 3 for more information.)

Vitamin E

DAILY DOSE

Tier 1 and Tier 2: *200 IU daily*
Tier 3: *400 IU daily*

Always buy d-alpha tocopherol and not dl-alpha tocopherol, since the latter is synthetic and is far less biologically active.

Vitamin E is a true "smart pill." Smart people take vitamin E to prevent brain aging, and people who take vitamin E stay smarter as they age. Researchers at the Chicago Health and Aging Project reported that older adults (65 plus) who consumed the highest levels of vitamin E from food or supplements had the mental function of people *eight to ten years younger* as compared to those who consumed little vitamin E. In other words, taking vitamin E can buy you a decade's worth of brain power. That's why I think that *everyone* should take this powerful antioxidant every day.

Vitamin E is a fat-soluble vitamin, which means that it can get into parts of the cell—notably the cell membrane—that are not accessible to other antioxidants. Since the brain consists of more than 60 percent fat, vitamin E plays a vital role in protecting brain cells from free radical attack. Vitamin E also inhibits the biological pathway that triggers inflammation, which is a causal, if not an aggravating, factor in most chronic diseases, including those that cause brain degeneration.

The evidence supporting the role of vitamin E as a powerful protector of the brain is overwhelming. A 1997 study published in the *New England Journal of Medicine* reported that vitamin E worked *better* than a commonly prescribed drug in slowing down the progression of Alzheimer's disease. In this landmark study, 341 patients with early Alzheimer's disease of moderate severity were given either 1,000 IU of vitamin E daily, Selegiline (a monoamine oxidase inhibitor), a combination of vitamin E and Selegiline, or a placebo for two years. After two years, patients taking vitamin E alone had a 53 percent lower risk of reaching the most severe stage of Alzheimer's than those taking the placebo. Those taking the drug alone had a 43 percent reduced risk, and those taking the drug and vitamin had 31 percent reduced risk. Vitamin E is the clear winner.

Researchers at Johns Hopkins University in Baltimore and other institutions analyzed the diet and health data from the famous Baltimore Longitudinal Study of Aging. They focused on 579 participants, all 60 years of age or older. During a nine-month period, 10 percent of the subjects developed Alzheimer's disease. Those who consumed the highest amount of vitamin E in their diets were at the lowest risk of developing Alzheimer's.

Even if you're not worried about getting Alzheimer's disease right now, keep in mind that if you live long enough, you have a 50 percent likelihood of getting this brain-debilitating disease. Taking a vitamin E supplement daily now will not only keep your brain younger, but you will significantly reduce your risk for Alzheimer's disease.

Folic Acid

DAILY DOSE

Tier 1: *400 mcg as part of a B-complex supplement*
Tier 2, Tier 3: *take an additional 400 mcg for a total of 800 mcg daily*

If you are moody or forgetful or have been diagnosed with depression, you may be lacking in folic acid, a vital B vitamin. And even if you're not suffering from any of these symptoms now, low levels of folic acid will vastly increase the risk that one day you will be! Numerous studies have linked mood disorder, memory problems, apathy, disorientation, and difficulties concentrating with folic acid deficiency. This deficiency is fairly common among the elderly, but women who take birth control pills and people over 60 are also at risk of low folic acid levels. Along with vitamin B12, folic acid helps to control homocysteine, the amino acid produced by the body that, if elevated, can cause inflammation and damage blood vessels in the brain. In fact, studies have shown that one-third of all people suffering from clinical depression have either elevated homocysteine levels or abnormally low folic acid or B12 levels. Many drugs deplete the body of B vitamins, including folic acid, which can send homocysteine levels soaring (see chapter 4 for a list of nutrient-depleting drugs). If you are taking any of these drugs, you must take additional B vitamins to compensate for the loss.

When you learn the facts about elevated homocysteine, you'll really be in a bad mood. It can threaten your life. High levels of homocysteine increases the risk of stroke, Alzheimer's disease, and heart disease. I believe that everyone should have their homocysteine levels checked as part of their annual physical (see chapter 10). The problem is relatively easy to correct by taking the right supplements, and this can spare you a lifetime of misery.

Folic acid is found in green leafy vegetables, dried beans, orange juice, peanuts, wheat germ, and fortified breakfast cereals. It can be destroyed by microwave cooking, overcooking food, exposure to sunlight, or overprocessing food. Due to its fragile nature, unless you take a folic acid supplement, you really can't be sure that you are getting enough of this vitamin in your diet.

That's why I recommend folic acid for everyone. It's included in the B-complex vitamin in Tier 1, and additional folic acid supplements are recommended for those following the Tier 2 and Tier 3 regimens.

Ginkgo Biloba

DAILY DOSE

Tier 3: *60 mg*

If you are following Tier 1 or Tier 2, you don't need ginkgo biloba, but if you are Tier 3, I believe this supplement can make a real difference in your mental performance and the quality of your life. I recommend ginkgo biloba to my Tier 3 patients who have moderate to severe memory problems and are at high risk for Alzheimer's disease or dementia. Scores of studies have confirmed that ginkgo can improve blood flow to the brain as well as enhance mental function and memory, but its positive effect appears to be most noticeable in people who already have cognitive problems. Ginkgo biloba has been used in Europe for decades as a prescription drug to treat problems related to brain function, heart disease, and other circulatory problems. Remarkably, it can improve mental function in people already diagnosed with dementia, for which there are virtually no effective drug therapies. In a landmark placebo-controlled, double-blind study published in the *Journal of the American Medical Association* (*JAMA*), ginkgo extract was tested on patients suffering from dementia caused by stroke or Alzheimer's disease. Of the 137 patients who completed the study, about 30 percent of those taking a 120-milligram capsule of ginkgo daily showed improvements on tests of reasoning, memory, and behavior, as compared to the placebo users. Considering that there are no prescription drugs that can achieve anywhere near this result, it is an important finding.

Ginkgo biloba is one of the most extensively studied nutritional supplements in the world. It is also one of the oldest herbal therapies, used by Chinese healers for thousands of years. Ginkgo's survival is probably due to its abundance of antioxidant flavonoids, which have fortified it against the environmental stressors that can shorten the lifespan of less resilient species. The same flavonoids that give ginkgo its strength are of great benefit to humans.

An important question is whether or not ginkgo biloba can improve mental function in normal adults. The results are not as clearcut and are more controversial. In one study, researchers looked at the effect of ginkgo supplementation on older adults who did not appear to have any cognitive problems. Participants who took 180 milligrams of ginkgo extract for six weeks showed significant improvement on several mental function tests as compared to placebo takers. Interestingly, more ginkgo takers said that their memory had improved by the end of the study as compared to the placebo group. But recently ginkgo biloba received negative publicity because of another study published in *JAMA* that found no improvement in mental function among healthy, elderly people taking 120 milligrams of ginkgo for four weeks. This study is one of the few that did not find any benefit in taking ginkgo biloba in terms of improved mental performance. I can't explain why the results of this study are so different from scores of other studies, but I don't take it too seriously. I have never regarded ginkgo biloba as a "smart pill"; rather, I feel that, like other antioxidants, it protects the brain from damage that could impair mental function down the road. Studies of healthy people are often trickier to perform than studies of people who have obvious deficits. It's more difficult to show an improvement in mental function among people who don't have a problem to begin with—sometimes changes are very subtle and hard to document on the standard mental function tests that are designed for people who have problems. There's no dispute that ginkgo biloba improves the flow of blood and oxygen to the brain and that it helps patients suffering from dementia. But keep in mind that even in the studies in which ginkgo biloba–takers with dementia showed improvement, they were hardly cured of their problem. They were still forced to live with severe cognitive impairments that interfered with their quality of life. The bottom line is: the Better Brain program will prevent these problems from taking hold in the first place!

N-Acetyl-Cysteine (NAC)

DAILY DOSE

Tier 2: *400 mg*
Tier 3: *400 mg taken twice daily for a total of 800 mg*

NAC is one of the few substances that, when taken orally, can raise blood levels of glutathione, an antioxidant produced by the body that is essential for the health of your brain and nervous system. Glutathione protects your brain against free radical attack and inflammation, which can destroy brain cells and disrupt brain function. The free radical–inflammation cycle is responsible for the mild memory and performance problems that start in midlife and lead to serious brain disorders decades later. If you don't have enough glutathione, the aging process will hit your brain faster and harder. Glutathione production declines with age, and it's very difficult to replenish it. Oral glutathione supplements are poorly absorbed by the body, but NAC can give the body a much-needed glutathione boost. (Acetaminophen can deplete your supply of glutathione. See chapter 4 for a list of drugs that can lower your glutathione level. If you take any of these drugs, be sure to take the NAC supplements recommended in chapter 4.)

If you are worried about exposure to toxic metals, NAC can help rid your body of heavy metal "pollution" and can chelate metals such as lead and mercury, both of which in high concentrations in the body can cause nerve damage. Exposure to high levels of mercury in particular can damage brain cells, resulting in severe cognitive problems. Many people are unknowingly exposed to heavy metals (see chapter 8). For example, mercury is found in high concentration in fish such as tuna and king mackerel. And some physicians believe that long-term exposure to mercury from "silver" in your teeth can also cause toxic mercury buildup in the body. A chelating agent such as NAC may help prevent long-term damage from exposure to this brain-toxic heavy metal.

If Alzheimer's disease runs in your family, here's another reason to take NAC: to keep your glutathione levels high. Glutathione detoxifies a free radical found in the brain that is associated with the production of amyloid plaques, the telltale sign of Alzheimer's disease. This is an important example of how an antioxidant can stop the downward spiral of brain aging in its tracks, and why it is so important to control free radicals.

Phosphatidylserine (PS)

DAILY DOSE

Tier 2: *100 mg daily*
Tier 3: *100 mg twice daily*

Would you like to make your brain 12 years younger? That's the promise of phosphatidylserine, called PS for short, a fatlike substance that concentrates in brain cell membranes. According to a recent study, PS can restore memory and improve cognitive performance in people suffering from age-related memory loss. One study, reported in the *Journal of Neurology* by researcher Thomas Crook, involved 149 healthy men and women age 50 to 70 who were diagnosed with "normal" age-associated memory impairment. The men and women were given either 100 milligrams of PS daily for 12 weeks or a placebo. At the end of the study, those taking PS showed marked improvements in their memory and were better able to recall telephone numbers, memorize paragraphs, and find misplaced objects. They were also better able to concentrate than the placebo-takers. According to Dr. Crook, PS supplementation gave patients back an average of 12 years in terms of their mental function. Interestingly, subjects who began with the most severe impairment showed the best improvements.

Why would taking PS supplements improve memory? PS is essential for healthy cell membranes. Without well-functioning cell membranes, your brain cells cannot communicate properly with each other, which means you will have difficulty learning new information or retrieving old information. When you can't place the face with the name and when you can't remember where you just put your keys, it's because your cells are not communicating with each other as well as they used to. PS is also essential for energy production within brain cells. The mitochondrial membrane, the actual site within the cell where fuel is transformed into energy, contains high amounts of PS. Without enough of this key brain nutrient, cellular energy production could be compromised, and without enough energy, your brain simply can't work as well.

PS production declines with age, which is why many of us need to take supplements.

Vinpocetine

DAILY DOSE

Tier 2 and Tier 3: *5 mg twice daily, for a total of 10 mg*

Vinpocetine isn't for everyone, it's for people who have high homocysteine levels and/or a history of heart disease or vascular dementia, that is, senility caused by impaired blood flow within the brain. Vinpocetine helps improve the flow of blood, oxygen, and vital nutrients to brain cells. In my practice, I have seen vast improvement in patients using this supplement; practically overnight they become more lucid and are sharper and more on the ball. (It should not be used if you are taking a blood thinner, such as Coumadin.) Vinpocetine is an extract of the periwinkle plant, *Vinca minor,* the same plant that has given us potent cancer treatments for childhood leukemia. For more than two decades, vinpocetine has been used in Europe and Japan to treat stroke victims and people suffering from dementia due to impaired blood circulation to the brain. More than 50 clinical studies have documented that vinpocetine can improve blood flow to the brain, promote better oxygen utilization, increase energy production in the brain, and help prevent blood clots. It is also a potent antioxidant. Vinpocetine became available over the counter in the United States in the late 1990s.

I have found vinpocetine to be extremely effective for my patients with vascular dementia, and there is a long list of studies that confirm what I have seen in my practice.

As far back as 1985, Japanese researchers described a "slight to moderate" improvement in two-thirds of all stroke patients receiving vinpocetine. This was a remarkable finding, considering the fact that brain injury due to stroke was believed to be irreversible, an erroneous concept still held by many unenlightened physicians. In a 1987 study published in the *Journal of the American Geriatric Society,* Italian researchers reported that when elderly patients with chronic cerebral dysfunction (vascular dementia) were given vinpocetine or a placebo, those taking vinpocetine scored "consistently better on all evaluations of effectiveness of treatment including measurements on the Clinical Global Impression (CGI) scale, and the Mini Mental Status Questionnaire (MMSQ). There were no serious side effects related to the treatment drug."

In a 1991 double-blind, placebo-controlled study in Germany, 203 patients were given either a daily dose of 30 milligrams of vinpocetine, 60 milligrams of vinpocetine, or a placebo for 16 weeks. Once again the vinpocetine group fared significantly better than the placebo group. "Patients treated with vinpocetine (both 30 mg and 60 mg/day) scored consistently better than the placebo group on both the physician's clinical global improvement ratings and the SKT, a short cognitive performance test for assessment of memory and attention."

At the Perlmutter Health Center, we do not recommend vinpocetine to patients taking blood-thinning medication (except for aspirin).

Your Supplement Regimen

The Better Brain program consists of three different supplement regimens, Tier 1, Tier 2 and Tier 3, which are based on the results of your Brain Audit. In the following section, I answer some of the most commonly asked questions about supplements. You may be familiar with supplements and know precisely what they are, where to get them, and how to use them. If so, you can skip this section and go right to the right program for you.

If I eat well, do I still need to take supplements?
The answer is yes. Unfortunately, modern food-processing techniques have depleted the food supply of many vital nutrients (such as B vitamins, vitamins C and E, and the essential fatty acids) that are critical for health. The storage and shipping of food, especially fruits and vegetables, can also sap it of its nutrients. If you don't supplement your diet with these missing nutrients, it is easy to fall short of them and suffer serious health consequences. Moreover, the body's production of key antioxidants (such as Co-Q10, alpha lipoic acid, and glutathione) slows down with age, which leaves us more vulnerable to free radical attack. In addition, as noted in chapter 4, many commonly used over-the-counter and prescription medicines also sap the body of important vitamins. Taking supplements can fill the nutritional gap and help maintain optimal mental and physical health.

What do the doses mean?
Supplements are micronutrients; you need to ingest only a small amount to yield a positive effect. Most supplements are sold in doses of micrograms

(1 millionth of a gram), milligrams (1 thousandth of a gram), or grams. The exception are fat-soluble vitamins (A, D, E, and K) which may be sold in International Units (IU). Basically, 1 IU is roughly equal to 1 milligram.

Please stick to my recommended doses, which are both safe and effective. Some supplements can be toxic if taken in extremely high doses.

Where should I buy supplements?

Supplements can be purchased at health food stores, pharmacies, discount stores, supermarkets, from mail order catalogues, and on the Internet. Some excellent brands of supplements are only available at doctor's offices. In Appendix 3 you will find a list of reputable distributors. In the interest of full disclosure, I mention here that I have designed a brand of brain supplements, called Brain Sustain and Brain Sustain Neuroactives (also listed in Appendix 3). I am presenting the supplement plan in generic terms so you can purchase your supplements wherever it is most economical and convenient for you.

Which brands are the best?

There are hundreds of brands on the market; my advice is to stick to reputable brands that you know. Safety should come first. Look for products that come in tamper-proof packages with an expiration date on the label. Try to find a product that has a quality-control number on the package; that way, if there is any problem the manufacturer can quickly recall a tainted product. If you have a choice, select manufacturers that offer a guaranteed-potency product, which means the supplement contains the right amount of the active ingredient. In particular, studies of many herbal products have shown that many contain little, if any, active ingredient. Look for products that are pharmaceutical grade, which are of the highest quality and free of impurities.

Better quality products may be a bit more expensive than the "bargain basement" brands, but they are well worth the difference. Manufacturers who provide safer packaging and guarantee the potency of their product have a higher overhead and will have to pass the cost on to consumers. But if you buy a cheap product that does not contain what it should, you're just throwing your money out.

How should I store supplements?

Most supplements should be stored in a cool, dry place away from direct light or heat. Some manufacturers may tell you to refrigerate a product (such as flaxseed or fish oil supplements) after it is open, so be sure to read the label carefully.

When should I take my supplements?

Take all supplements with meals. Most of you will be taking supplements twice a day, the first dose with breakfast and the second dose with lunch or dinner. Most supplements are best absorbed when taken with food. To make it easy for yourself, spend a few minutes each week organizing your daily supplements for the entire week. Put your daily dose into a pill box divided into the days of the week, or in a small plastic bag. (Use one bag for each dose.) Plastic bags work well if you need to carry a midday dose to work with you.

Which forms of supplements are most effective?

Supplements come in many different forms, from pills to capsules, liquid extracts, or powders. Choose the one that is easiest for you to use.

What can I do if taking supplements upsets my stomach?

Although most people tolerate supplements fairly well, some do find that they suffer stomach upset after taking their supplements. Very often your problem is not with all supplements but with one supplement that is giving you grief. To find out which one it is, you need to discontinue taking all supplements for three days. Then resume taking your supplements one supplement at a time. If you don't feel any stomach distress in a day or two, add on an additional supplement the next day. If your symptoms resume after taking a particular supplement, drop it from your regimen. At the end of this exercise, you should know which supplements you can tolerate and which you cannot.

If you're pregnant or have kidney disease, do not take any supplements without consulting with your doctor.

Supplements and Prescription Medicine

If you are taking a prescription medicine, it's always wise to tell your doctor about any supplements that you may also be taking. In most cases it's fine, even desirable, to take supplements, especially in the case of drugs that deplete the body of specific nutrients. If you follow my prescribed doses for supplements, you should not experience any adverse interaction with prescription medicine, but there are two exceptions.

Vinpocetine. Do not take vinpocetine if you are taking a prescription blood thinner like Coumadin (warfarin) because it is also a blood thinner. (Please note that if you are scheduled to have surgery, you need to tell your doctor if you are taking any natural or prescription blood thinners, notably vitamin E and ginkgo biloba. You may have to discontinue them prior to surgery to avoid excess bleeding.)

Vitamin C. Vitamin C supplements can enhance the action of some antidiabetic drugs, which means that you may need to adjust your dose. If you are taking vitamin C and are under treatment for diabetes, talk to your doctor.

Finally, some antioxidants should be avoided during certain chemotherapy regimens, since many chemotherapy drugs work by increasing free radical production to kill cancer cells. To be on the safe side, if you're taking any prescription drugs, check with your physician or pharmacist for any potential drug interactions or negative side effects.

Tier 1: Prevention and Maintenance

SUPPLEMENT	A.M.	P.M.
DHA	300 mg	
Co-Q10	30 mg	
Vitamin E	200 IU (d-alpha, *not* dl-alpha form)	
Vitamin C	200 mg	

Vitamin B-complex supplement*		
B1 (thiamine)	50 mg	
B3 (niacin as niacinamide)	50 mg	
B6 (pyridoxine)	50 mg	
Folic acid	400 mcg	
B12 (cobalamin)	500 mcg	

*Look for a B-complex supplement containing several B vitamins in one capsule or pill.

Tier 2: Prevention, Repair, and Enhancement		
SUPPLEMENT	A.M.	P.M.
DHA	300 mg	
Co-Q10	60 mg	
Vitamin E	200 IU	
Vitamin C	200 mg	200 mg
Alpha lipoic acid	80 mg	
N-acetyl cysteine (NAC)	400 mg	
Acetyl L-carnitine	400 mg	
Phosphatidylserine	100 mg	
Vinpocetine*	5 mg	5 mg
Vitamin B-complex supplement:**		
B1 (thiamine)	50 mg	
B3 (niacin as niacinamide)	50 mg	
B6 (pyridoxine)	50 mg	
Folic acid	400 mcg	400 mcg***
B12 (Cobalamin)	500 mcg	500 mcg***

*Vinpocetine is required only for people with high homocysteine or a history of vascular dementia or coronary artery disease. It should be avoided by people taking Coumadin (warfarin), a blood thinner.

**Look for a B-complex supplement containing several B vitamins in one capsule or pill.

***In addition to your B complex in the A.M., take an additional 500 mcg of B12 and 400 mcg of folic acid in the P.M.

Tier 3: Recovery and Enhancement		
SUPPLEMENT	A.M.	P.M.
DHA	300 mg	300 mg
Co-Q10	100 mg	100 mg
Vitamin E	400 IU	
Vitamin C	200 mg	200 mg
Alpha lipoic acid	200 mg	
N-acetyl-cysteine (NAC)	400 mg	400 mg
Acetyl-L-carnitine	400 mg	400 mg
Phosphatidylserine	100 mg	100 mg
Ginkgo biloba	60 mg	
Vitamin D	400 IU	
Vinpocetine*	5 mg	5 mg
Vitamin B-complex supplement:**		
B1 (thiamine)	50 mg	
B3 (niacin as niacinamide)	50 mg	
B6 (pyridoxine)	50 mg	
Folic acid	400 mcg	400 mcg***
B12 (Cobalamin)	500 mcg	500 mcg***

*Vinpocetine is required only for people with high homocysteine or a history of vascular dementia or coronary artery disease. It should be avoided by people taking Coumadin, a blood thinner.

**Look for a B-complex supplement containing several B vitamins in one capsule or pill.

***In addition to your B complex in the A.M., take an additional 500 mcg of B12 and 400 mcg of folic acid in the P.M.

Change Your Lifestyle, Change Your Brain

ELIZABETH, 47, is an attorney who came to see me because she felt that her mental performance was slipping, and it was beginning to affect her ability to function at work. She had long prided herself on her ability to remember names, dates, and other details of importance to her clients but was now having difficulty doing so. She was having trouble coming up with the right words to express her thoughts (a real problem for a trial lawyer, who must think on her feet), and, much to her chagrin, she was finding it harder and harder to stay alert during late afternoon meetings.

"Why is this happening to me now?" Elizabeth asked. "I know I work long hours, but I'm not doing anything that I haven't done before. Now I can't seem to keep up with it all."

I asked Elizabeth some questions about her habits and lifestyle, similar to the questions on the Brain Audit. After a few answers it became obvious to me why Elizabeth was not functioning at her mental or physical peak.

Elizabeth's lifestyle was a road map to a stressed-out, tired, poorly functioning brain. She worked 14 hours a day, including at least one day on the weekend, rarely got more than five to six hours of sleep a night, and rarely exercised (even though there was a gym right in her office building). When I asked her if she had any hobbies, she snapped back, "Who's got the time?"

Over my 20 years of medical practice, I've seen hundreds of patients like Elizabeth, and reaching a diagnosis was easy. I told her that I believed her problem was the result of her high-pressure, "must get it done *right now*" lifestyle. Of course, I would run all the usual medical tests to rule out any hidden serious conditions, but I doubted that she had one (and I was right, she passed her medical tests with flying colors). Although Elizabeth might have been able to get away with running herself into the ground when she was younger, it wasn't going to work for her anymore. I told Elizabeth a bit about how free radicals were beginning to overwhelm her brain's antioxidant defense system and how those energy-producing mitochondria in her brain cells were starting to slow down. I explained to her how the shortage of energy in her brain cells could make it harder for her to learn new things or concentrate, and how lack of sleep and constant stress could eat away at her brain cells by not giving her brain any time off for rest or repair. I warned Elizabeth that if she continued to make constant demands on her brain without giving it any down time, she would quickly lose the mental edge that made her a great lawyer.

Based on her symptoms and lifestyle, I put Elizabeth on the Tier 2 supplement program and the Better Brain meal plan (see chapter 5). I also gave Elizabeth my lifestyle prescription for peak brain performance and good mental health. I call it "The Three Rs": rest, relaxation, and recreation. This prescription is easy to follow but of vital importance for a well-functioning brain.

Three days later, a delighted Elizabeth called me to thank me for saving her career. For the first time in a long time, she felt articulate and alert and that she was well worth her fee of $200 an hour. I know my program works, but it does take a little more time than three days to kick in! I knew Elizabeth's amazing transformation had come about simply because, for the first time in years, she had slept for eight hours straight for three nights in a row. To Elizabeth's brain all that sleep was the equivalent of going on an extended vacation! I also knew that the longer

Elizabeth stayed on the Better Brain program, the more alert and energized she would feel.

Get Enough Rest

If you're not getting between seven and eight hours of sleep most nights, you are probably not performing at your mental peak, and you are putting the health of your brain at risk. According to a 2002 poll taken by the National Sleep Foundation, fewer and fewer people these days get the seven to eight hours of sleep a night required for optimal mental and physical performance. On average, we sleep 20 percent less than we did one hundred years ago, when nine and a half hours sleep a night was the norm. These were the days before electric lights made it possible to ignore the dark, or when restaurants served patrons 24 hours a day and late-night talk shows ruled the airwaves.

Although we choose to sleep less, it doesn't mean that we are making the right choice. Chronically sleep-deprived people may not feel particularly tired, yet they often experience deficits in cognitive function. Even a mild sleep deficit—getting six hours of sleep for a few nights instead of eight hours—can cause notable declines in reaction time and performance on standard mental function tests. What are the signs of sleep deprivation? If you are sleep deprived, you'll have a more difficult time learning a new task, you will find it harder to juggle several tasks simultaneously, and in sum, you will not be working up to your full potential.

Missing sleep doesn't just make you feel lousy, it can cause long-term damage to your brain cells. Sleep is not a luxury; it serves an important purpose. Most systems wind down while we are sleeping, giving the body time to rest and refuel. As the body shifts into low gear, heart rate and blood pressure drop, and metabolism slows down. Sleep is a time when our cells can concentrate on repairing themselves and making new cells. For your brain, this is a particularly important time. While you are sleeping, levels of neurotransmitters (the chemicals that help you think, remember, and maintain a good mood) are normalized, and energy production by neurons is slowed down so that your brain cells can clean out waste products (like free radicals) that can accumulate in the brain. If you don't get enough sleep, this vital repair work will not get done.

Sleep is also a time when your brain processes information. You know the old expression "I'm going to sleep on it"? Some studies suggest that sleeping—or actually dreaming—may be a time when people subconsciously solve problems, which may enhance the ability to learn new material.

Losing even one night's sleep has a marked physiological effect on the body. The disease-fighting cells of your immune system are weakened, leaving you more vulnerable to infection. Your body pumps out higher levels of stress hormones, which cause a spike in blood sugar levels. In fact, a recent study suggests that healthy young people who get less than six and a half hours of sleep a night were at greater risk of insulin resistance, a prediabetic condition, as compared to people their age who got between seven and a half and eight and a half hours of sleep a night. This is particularly alarming because diabetes can accelerate brain aging and make you more vulnerable to neurological disease. Lack of sleep is also linked to mood disorders such as depression, irritability, and anxiety.

Half the battle is simply making up your mind that sleep is important to your health and well-being, and showing a willingness to set aside the hours to get enough sleep. Once Elizabeth agreed to go to bed two hours earlier at night, she quickly adjusted to the new sleep schedule. For many people, simply making the commitment to sleep is all it takes to improve sleeping habits. Other people, however, may have problems that prevent them from getting a good night's sleep.

About 40 million Americans suffer from sleep disorders; that is, they report difficulty sleeping at night. Many resort to taking medication to help themselves sleep, but these drugs can be habit-forming and should never be used for more than a week or two. In my experience, I have found that sleep disorders are often caused by unrelated problems, such as excess caffeine consumption, smoking (nicotine is a stimulant), menopausal hormonal swings, or even depression. Once these problems are corrected, normal sleeping patterns can be restored. If you have difficulty sleeping—that is, if you can't fall asleep within 15 to 30 minutes at night, if you wake up frequently during the night, or don't wake up feeling refreshed—please review the following.

Do You Have an Underlying Medical Problem? If you are in discomfort due to arthritis or back pain, it could be keeping you up at night. Other medical conditions, such as thyroid disorders, liver problems, and neurological problems, could also interfere with sleep. So can numerous medications used to treat these and other problems. If you are tossing and turning at night, be sure to get a complete physical examination to rule out any underlying medical condition, or medicine you may be taking for it, as the cause.

Are You Overcaffeinated? Caffeine is a powerful stimulant, and that's why many people rely on that morning cup of coffee to get them going. Although some people can tolerate large amounts of caffeine, caffeine-sensitive people may find that it makes them jittery and can interfere with sleep. Unfortunately, it's difficult to avoid caffeine. In addition to coffee and tea, caffeine is in numerous products, ranging from colas (unless they are labeled caffeine free) to soft drinks to chocolate and over-the-counter and prescription painkillers. Even one cup of coffee in the morning can cause insomnia in some caffeine-sensitive people. If you find that you are up tossing and turning at night, and you are a caffeine user, try cutting back or cutting out caffeine altogether. I have seen many cases of insomnia cured simply by eliminating caffeine from the diet. If you are a heavy caffeine user, you may need to gradually decrease your intake of caffeine products or you may suffer a headache for a few days as your body "decaffeinates."

No Smoking at Night. I know that people who smoke say that it relaxes them, but in reality nicotine is a stimulant. It's best not to smoke at all, but if you do, refrain from smoking for least two hours before bedtime.

Pass on the Nightcap. Alcohol may knock you out quickly, but it can interfere with the deepest phases of sleep and cause frequent nighttime awakenings.

Exercise Helps . . . to a Point. Lack of exercise is a major cause of insomnia, but too much exercise too close to bedtime can also cause insomnia. A high-powered gym workout can increase your heart rate and rev you up for action. It is not a good idea to do a serious workout before bedtime. On the other hand, gentle stretching or a walk outdoors is a very good way to wind down before bedtime.

Is Stress Keeping You Up? If you find that you are up at night fretting over the day's events or worried about what's next, read over the following section.

Natural Solutions. Some herbal teas have natural sedative properties—they'll make you sleepy without making you feel drugged, and unlike sleeping pills, they are nonaddictive. Try drinking a cup of chamomile, skullcap, lemon balm, or passionflower tea before bedtime. (If you have a ragweed allergy, skip the herbal tea.) Melatonin, a hormone sold over the counter in health food stores and pharmacies, can help reestablish normal sleeping patterns. Melatonin is produced by the pineal gland, a pea-sized structure embedded deep within our brains. It is the hormone that regulates the sleep-wake cycle. As we age, our production of melatonin declines, which many scientists believe is the reason why sleep disorders are so common among the elderly. Try taking 3–9 milligrams of melatonin an hour before bedtime. Start with the lower dose, and if it doesn't work, take an additional 3 milligrams, and if you still need more, take another 3 milligrams until you reach your total of 9 milligrams. After a few nights, your own natural sleep cycle may be restored, and you may only need to use it on occasion. I have found that melatonin is particularly good for women in midlife who are suffering from sleep disturbances due to menopausal symptoms. Melatonin is also a powerful antioxidant. I recommend it to patients with Alzheimer's disease because they often have disrupted sleep patterns, and are typically deficient in this hormone.

Relax: Control Stress Before It Controls You

We experience stressful situations every day, whether it's running for a commuter train, or rushing a child off to school, or trying to fit 25 hours' worth of activities into a 24-hour day. Stress per se is not all bad. In small doses, some types of stress can be exhilarating, even beneficial, such as the stress you feel when you strive to meet a new physical or mental challenge (i.e., mountain climbing or solving a problem at work, even doing the brain workout in chapter 9), or acquire a new skill (i.e., learning to speak a new language or operate a new computer system.) But chronic, unrelenting, uninvited stress is a different matter. This kind of stress can be toxic to your brain and cause physiological changes in the brain that can profoundly affect mental performance and accelerate brain degeneration.

When you are under stress of any kind, your body produces corticosteroids, special hormones that trigger the ancient "fight-or-flight" response that we inherited from our earliest human ancestors. These hormones rev the body up for action so that we can escape a predator or hunt down our prey. Under ideal conditions, production of these hormones is shut off quickly, and their effects dissipate. Things don't always work the way they are supposed to work. As you get older, it is more difficult for your body to dispose of stress hormones. As a result, they linger in your body longer than nature intended. If you are under chronic stress—for example, you're in the midst of a divorce or working in a high-pressure environment—you are producing high levels of corticosteroids and are bombarding your brain cells with these powerful chemicals for extended periods of time. In high doses, corticosteroids can damage your brain. First, corticosteroids can promote the formation of free radicals and inflammation. This in turn will damage the energy-producing part of the brain cell—the mitochondria—which will leave you with less energy to run your brain or to clean up free radicals. This will accelerate brain degeneration. Second, corticosteroids are particularly toxic to cells in the sensitive hippocampus area, the brain's memory center. In fact, some scientists believe that age-related memory loss may be a result of a lifetime of exposure to stress hormones. Third, chronic exposure to stress hormones can disrupt the production of neurotransmitters in your brain such as serotonin, acetylcholine, and dopamine, which can affect your mood, your cognitive function, or both. Too much stress can make you depressed, irritable, and forgetful. When you are stressed out, it is very difficult to summon the mental energy required to learn new material, concentrate on projects, or stay on task. And when you don't perform at your own level of expectation, you become even more stressed out.

Stress hormones also produce measurable and specific biochemical changes in your body that increase your risk of brain aging and neurological disease. For example, stress can raise the levels of homocysteine, an amino acid naturally produced by your body. This is not a good thing: high homocysteine can impair mental performance, as well as increase the risk for Alzheimer's disease and stroke. Every time you are exposed to stress, your homocysteine levels edge upward. A study of middle-aged women performed at Ohio State University found that brief periods of stress raised their homocysteine levels from an average of 5.8 to 6.2 micromoles per liter. (Normal is considered anything under 9 micromoles per liter.) Although this increase did not kick up homocysteine to unhealthy levels for these women, it could for people who start out with higher levels. Moreover, this study investigated the impact of one stressful event, not the effect of chronic, unrelenting stress, as is experienced by many people. (Do you know your homocysteine level? You should. See chapter 10 for more information.)

A word on recreational drugs. People use recreational drugs to make them feel better, happier, or high. Any drug that produces a marked change in mood will affect the health of your brain. Even antidepressants, which are liberally prescribed by physicians, produced by reputable pharmaceutical companies, and carefully dosed, have their downsides. Recreational drugs that are not regulated, can be tainted with dangerous chemicals, and are haphazardly dosed are an extraordinarily risky proposition, and I urge you not to even consider using them. I have treated many "smart" people who thought that they could handle recreational drugs, only to wake up one morning and find that a piece of their memory was missing. It's often impossible to help these individuals. Like painkillers, mood-altering drugs simply mask the symptoms of too much stress, unhappiness, loneliness, and anxiety. If you regularly use these drugs, you need to consider more constructive outlets for your mental distress. As a neurologist, I frequently treat emotional problems, and I often refer patients to mental health professionals, particularly when a person is engaging in risky behavior that is jeopardizing the health of his or her brain.

Stress is a reality of twenty-first-century life, and there's no escaping it. Many things in life are out of our control. An aging parent suddenly falls ill, and you are responsible for her care, or the company that you work for goes bankrupt, or the IRS decides to audit you. All of these events can be stressful. Although you don't have the power to stop bad things from happening, you do have the power to control how you respond to them. You don't have to let stress get the better of you, and you certainly don't have to sit idly by while it puts your brain and body in jeopardy. If you are under a great deal of stress, you should take immediate action to "destress." I recommend specific relaxation techniques to my patients that I practice myself and that I will share with you here.

A word of caution: stress-management techniques work well for people suffering from standard, run-of-the-mill, "that's life" forms of stress. They don't work for everyone, and I don't recommend them for everyone. If your stress is rooted in a past trauma, such as physical or sexual abuse, you should not try to cope on your own. You need to seek professional counseling. The same is true if you are feeling depressed, suicidal, hopeless, and in general very unhappy. Self-treatment is not an option for you. You must get professional help.

For everyone else, the following techniques often do the job.

Exercise

If your stress is rooted in the everyday stressors—job stress, financial stress, family stress—taking time out from your day to "destress" can be extremely helpful. I have a heavy patient load, and I try to make myself available to my patients whenever they need me. I have two great kids, and a wonderful wife, but they also require attention. Sometimes I feel pulled in every direction. So what do I do to stay centered and sane? No matter how busy I am, I take a one-hour break every day at lunchtime and work out at the gym. Exercise is a great stress reliever—it burns up excess stress hormones quickly and effectively. It also reduces your risk of obesity, diabetes, heart disease, and some types of cancer. By taking this time out for me, whenever possible, I am better equipped to take care of others. I'm not saying that everyone should go to the gym to relieve stress—frankly, some people may find gyms to be very stressful. I do believe that everybody needs some activity that enables the letting go of daily stress and fortifies the spirit.

Exercise is also a great mood booster. I recently saw a study that concluded that moderate exercise worked as well as prescription antidepressants in treating depression. If you are moody and irritable, a regular exercise program can help to put you back into the right frame of mind.

There's yet another compelling reason to make exercise a regular part of your life. It really can *save your brain*. Just a modest exercise program—taking a 20-minute walk every day—can reduce your risk for Alzheimer's disease by 30 percent!

I want to stress that you don't have to become an athlete, but you do have to do at least 30 minutes of some kind of physical activity for five days each week. Ideally, you should do a combination of aerobic (cardiovascular exercise to improve heart function and circulation) and weight training (to make muscle.) If you have never exercised before, or need to be motivated, you may find it helpful to join an exercise class or hire a fitness trainer, at least until you feel confident enough to do it on your own. There is a huge variety of exercise venues to chose from, from basic gyms at your local Y to fancy health clubs to night classes at your local high school. Whatever you do, do something to burn calories and strengthen your muscles. If you can't bear the thought of a structured exercise class, you can use a stationary cycle or ride a bicycle outdoors (see box). Even a brisk walk every day can be beneficial and help improve muscle tone. Whatever you do, it's critical that you set aside a certain amount of time at least five days a week to do it. Your appointment with yourself is as important as your appointment with others. Don't break it!

Consider Yoga

Yoga is an ancient science that incorporates deep breathing techniques (called pranayamas) and gentle exercises (called asanas) to improve phys-

Get a Pedometer

The average person takes far fewer than 4,000 steps a day—well under 1 mile—which isn't nearly enough to burn calories or maintain health. The Centers for Disease Control recommends increasing that amount to at least 10,000 steps a day, which would be equivalent to walking about 2 miles. No one expects you to count your steps, but you can get a handy device called a pedometer that will track how much you walk. It costs less than $20, and you can clip it on your clothes. Once you begin counting your steps, chances are you'll be more likely to take the steps, walk instead of drive when you can, or embark on a daily walking regimen.

ical strength and flexibility and reinvigorate the spirit. The yoga breathing exercises revitalize the body and the mind. Yoga breathing decreases the activity of the sympathetic nervous system, which triggers the flight-or-fight response and the production of stress hormones. In particular, yoga breathing has a calming effect on the "worry center" in the brain stem called the locus ceruleus. These exercises leave you feeling calm and refreshed. Some forms of yoga include meditation, which can produce a deep relaxation response in people who know how to do it. Meditation helps to clear the mind and improves concentration and mental clarity.

In recent years, yoga has been growing in popularity by leaps and bounds. In addition to helping you destress, yoga can also provide an excellent physical workout and offers specific heath benefits. Certain forms of yoga can lower high blood pressure, relieve arthritis, and reduce the risk of heart disease. You should be able to find a class or yoga instructor in your area. Check at your local Y, church, high school adult learning center, or yoga center. There are some wonderful yoga classes on television, or you can rent or buy an instructional video or CD on yoga. See Appendix 3 for more information.

Don't Know How to Relax?

Many people have been wound up so tightly for so long that they simply don't know how to let go and relax. Fortunately, there's a world of wonderful tools that can show you how to destress and enhance relaxation. For example, there are scores of videos, tapes, and CDs on the market that teach specific stress reduction and relaxation techniques. For some specific recommendations, turn to Appendix 3.

Recreation

One of the questions on the Brain Audit was about whether or not you engage in any leisure activities. Doing something you enjoy outside of work not only makes life a richer experience but helps preserve brain function. In chapter 3, I explained that people who do some activity beyond work are at lower risk of developing Alzheimer's disease than people who don't. This fact was revealed in a study investigating the impact of lifestyle as a risk factor of Alzheimer's disease, but I have a hunch that it will prove to be true for virtually every illness that can be aggravated by stress. Whether it's painting, dancing, tennis, reading, or knitting, when you are concentrating on an activity that you enjoy, it's possible to shut out all the irritants that cause stress. You're focusing on the task at hand, and you're not preoccupied with your annoying coworker, or your sullen teenager, or your sick parent. You've given yourself a "time-out" from stress, and given your body a chance to recover and regroup.

Participating in a nonwork activity may also stimulate portions of your brain that would otherwise lie fallow. When you stimulate brain cells, you make new dendrites, the connections between neurons that are essential for the assimilation and processing of information. The more dendrites you have, the better your brain cells can communicate, the "smarter" you will be, and the longer your brain will last.

Wear a Protective Helmet
When Engaging in Sports

Attention, all bike riders, skiers, snowboarders, skateboarders, and skaters: in the event of an accident, wearing a protective helmet can possibly save your life. It can also reduce your risk of Parkinson's disease or other neurological problems stemming from a head injury.

Despite my Florida upbringing, I am an avid skier. I take my family skiing every year, and I insist that we all wear protective helmets. It amazes me that we are often the only ones on the slope wearing them. We are all experienced skiers and have pretty good control, but that's not the issue. The ski slopes are filled with inexperienced skiers and snowboarders who are frequently out of control, and if you get hit by one of them, it doesn't matter whether you are an Olympic-ranked skier or not, you are just as vulnerable to sustaining a permanent and severe head injury.

There are about 18,000 head injuries per year from skiing and snowboarding. According to the Consumer Product Safety Commission (CPSC), close to eight hundred of those head injuries could be prevented or reduced in severity by wearing a helmet. In 1999, the CPSC began recommending that skiers and snowboarders wear helmets.

Bicycle riding is an even more popular sport than skiing and snowboarding and is a major cause of head injury. More than half a million bicyclists are taken to the emergency room each year with injuries, and one in eight of them has a brain injury. Eight hundred bicyclists die in the United States each year, primarily from brain injuries. Half of the deaths are in children under 15. Wearing a bike helmet can reduce the risk of head injury by about 85 percent and the risk of brain injury by 88 percent. Many states have mandatory bicycle helmet laws for children, yet surveys show that they are often ignored. According to the CPSC, only 15 percent of children under age 15 wear helmets all or most of the time. As far as I'm concerned, adult bicyclists also need helmet protection, but there are no laws mandating them to wear them. Some states do not require motorcyclists to wear helmets, but, given the high speeds that you are traveling at and the generally poor traffic conditions on the roads these days, I think you would be ill advised not to.

Skating, on ice or on land, and skateboarding are other sports where there is a high risk of head injury. Kids and adults who engage in these activities should also wear protective helmets.

You don't need a different helmet for each activity. Purchase a multisport helmet that complies with CPSC guidelines (it will have a CPSC sticker on the inside.) There are several good ones on the market that can be purchased at bike shops and sporting goods stores. Helmets come in different styles and sizes, and you may have to try a few on to find the right one for you. Be sure to buy one that fits well and isn't too loose, or it could slide to the back of your head on impact and not provide adequate protection. For more information on how to select a helmet, check out the Bicycle Helmet Safety Institute website at www.bhsi.org.

Get the Toxins Out!

DO YOU LIKE to garden? Do you live in a quaint, old house? Do you use deodorant? Do you eat off trendy, ceramic plates? Do you walk around with your cell phone glued to your ear? If you answered yes to any of these questions, you could be exposing yourself to toxins that could kill brain cells, lower your child's IQ, make you senile, or make you sick.

We are exposed to neurotoxins every day that can damaging to brain and nerve cells. They are everywhere—in places where you would least expect them. They are in food (mercury in fish and pesticides on produce) and personal care products (aluminum in deodorants, shampoo, and skin creams,) and in the walls and plumbing of older homes (lead in paint and pipes). They are even in the air. Radio waves from TVs, computers, and cell phones can be harmful to nerve cells, which is why I refer to them as electronic toxins. What makes toxins "toxic"is the fact that they rev up free radical production and promote inflammation, which can spread everywhere, including the brain. Exposure to neurotoxins can speed up brain degeneration and accelerate brain aging.

The truth is, we live in a very toxic world, and we are constantly challenging our bodies with new toxins. These toxins are finding their way into our bodies and into our brain cells. In a recent study conducted at Mount Sinai Hospital in New York, blood and urine samples were taken from nine volunteers to check their levels of known toxins. The scientists were surprised to find that the nine participants had traces of 53 known carcinogens, 55 associated with birth defects, and *62 chemicals known to be toxic to the brain and nervous system.* Granted, these individual chemical toxins may be at levels that are deemed "safe" by government standards, but the truth is, we don't know the long-term health effects of low-level but constant exposure to a multitude of chemicals. Moreover, most of these chemicals did not even exist 75 years ago. And it is this new onslaught of toxic chemicals that represents the greatest threat to brain health in human history.

Our bodies have a way of disposing of chemical toxins—the detoxification system. The liver is responsible for cleaning up toxic chemicals found in the bloodstream. Glutathione, a critical antioxidant for the brain, is found in high concentrations in the liver. When glutathione encounters a toxin, it attaches to the compound and makes it water soluble, allowing it to be flushed out in urine. In theory, the body's detoxification process should keep toxins under control, but in reality, the proliferation of modern-day toxins may far exceed our ability to handle these brain poisons. In many cases, there may not enough glutathione on hand to do the job effectively, and low levels of toxins may gain entry into the brain.

It's impossible to eliminate exposure to all chemicals, and few of us would be willing to give up the modern conveniences that create toxic pollution. We can, however, try to minimize our exposure to the toxins that may pose the greatest risk to the brain and nervous system. I have identified six common toxins that I feel are the most worrisome in terms of mental function and brain health. These are pesticides, mercury, aluminum, lead, excitotoxins (food additives), and EMFs (radio waves). While you can't avoid contact with them completely, it's certainly possible to significantly reduce your exposure without compromising your lifestyle.

Say No to Pesticides

Of all the potential brain toxins in the environment, I consider pesticides to be the worst offenders. Why? By definition, pesticides are designed to kill living organisms. Although they are supposedly safe for humans in small doses, I believe that any exposure to pesticides can be dangerous. Pesticides target delicate nerve tissue (that's how they kill pests) and in the process promote free radical production and inflammation. Pesticide residue is stored in your fat cells and can remain in your body indefinitely. Today's fleeting exposure to a pesticide can be tomorrow's nightmare. There is a clear, documented link between exposure to pesticides and an increased risk for Parkinson's disease, ALS, and other neurological diseases. But that doesn't mean that if you don't get Parkinson's, pesticides have not taken their toll on your brain. The same insidious process that leads to all neurological disease begins with free radicals and inflammation. Anything that accelerates that process is going to have a direct effect on how well your brain functions today, as well as your ability to maintain your lifestyle into your later decades.

In recent years, some natural pesticides (i.e., rotenone, an extract from the roots of the derris plant in Asia and cube plant in South America, and pyrethrum, derived from the dried flower of the chrysanthemum plant) have been touted as safe alternatives to chemical pesticides. A closer study of these pesticides, however, revealed that they are also neurotoxins, and may increase the risk of Parkinson's disease. In fact, rotenone is frequently used in scientific research to produce Parkinson's in experimental animals! The bottom line is, there is no such thing as a safe pesticide.

Outdoor Pests

Fortunately, there are some excellent alternatives to commercial pesticides that I personally use to control pests inside and outside of my home. If you are concerned about protecting your lawn from pests, there may be better ways to do it than to saturate the ground with poisons.

At our home in south Florida, my family and I tend an organic garden from which we get fresh fruits and vegetables. Given the heat and humidity of Florida, keeping the local bug population from eating the garden is a constant challenge but not an impossible one. We use a variety of innovative and safe methods to control insects quite successfully. About five years ago, we introduced beneficial insects into our garden, who prey on the more destructive insects. There are many kinds of insects that can be used to control a particular pest. For example, since we grow tomatoes, we introduced trichogramma, a tiny wasp that attaches to the eggs of moths and butterflies that produce tomato fruit worms and caterpillars. No, the wasps did not attack us—they were too busy controlling crop-destroying insects. Granted, it's a little extra work not to use pesticides, but it is well worth the time in terms of protecting our health. You can purchase beneficial insects from companies that supply organic farmers and gardeners (see Appendix 3).

Certain plants can be used to protect each other. For example, basil helps repel flies and mosquitoes from tomato plants and improves the growth and flavor of tomatoes. We have also learned that you don't need herbicides to control weeds: all you need to do is simply pull them out of the ground, or you can use a nontoxic, natural herbicide made from corn

Herbal Insect Repellent

Long before the invention of chemical pesticides, farmers and gardeners used safer, nontoxic substances to control unwanted pests. Herbs such as garlic, pepper, and onion can be used to make an effective and safe insect repellent. (Place a teaspoon of hot pepper, an onion, and three garlic cloves in two quarts of water and heat to boiling. When it cools, pour the solution through a strainer and put it in a spray bottle.) This nontoxic but foul-tasting garlic-based insect spray can be used on roses, azaleas, and vegetables to kill bugs. You can make your own garlic-based insect spray or buy one of several commercial brands on the market (see Appendix 3 for more information).

gluten (see Appendix 3). If you are interested in pesticide-free gardening, I recommend that you subscribe to *Organic Gardening*, published by Rodale Press.

Mosquitos can also be controlled without declaring chemical warfare. Be vigilant about not allowing pools of water to gather in your backyard where mosquitoes can lay their eggs. Cover up when you go outdoors; wear long sleeves and long pants and shoes with socks in areas that are infested with mosquitos. Don't wear perfume or aftershave that can attract insects. Use a natural bug repellent such as Bite Blocker, a soy-based insecticide, or Burt's Bees Herbal Insect Repellent (see Appendix 3 for more brands). Any bug repellent containing citronella is also a safe choice. Oil of citronella repels mosquitoes, flies, fleas, and ticks. When used as directed, it's harmless to humans; it only works for up to two hours and must be reapplied. Products containing diethyl-m-toluamide (DEET) work longer, but DEET is a neurotoxin. Some scientists believe that DEET exposure has a role in causing "Gulf War syndrome," a disease characterized by severely abnormal brain function. Although DEET is considered safe when used as directed, we have never used it in my family and never will. When my son went on a two-week camping trip last summer, he never once used DEET; instead he relied on herb-based insecticides made from lemon grass oil, citronella oil, and rosemary oil, and he managed to avoid being bitten.

Indoor Pests

You can also control pests inside your home without risking the long-term health of your family. Bugs wander indoors looking for food and water. If you make your home inhospitable, they will go elsewhere. Don't allow food to be eaten throughout your house; confine eating to the kitchen or dining room. Keep your countertops and kitchen floor crumb free. Put all food away in covered containers; make sure the kitchen sink is wiped dry; and fix any leaky faucets. If ants are coming into your home, sprinkle some cayenne pepper, paprika, or boric acid at their point of entry. It will stop them dead in their tracks. If cockroaches are your problem, try this old-fashioned solution. Mix 6 ounces boric acid,

4 ounces sugar, and 8 ounces flour in a bowl. Spread the mixture on the floor and on countertops, especially around the cracks and crevices where roaches tend to hide. Do not allow the boric acid mixture to come in contact with food (although it's nontoxic, that doesn't mean it can't make you sick). Reapply a fresh mixture after four days, and then again after two weeks. It should help control the problem. (Keep pets and children away from the boric acid mixture.) For more information on natural pest control, see Appendix 3.

Getting Pesticides off Your Plate

Since most commercially sold food is treated with pesticides, I recommend that you choose organic products when you have a choice. Organic produce is grown without pesticides or other chemical additives. For more information on organic produce, see Appendix 3.

Get the Mercury off Your Plate

Remember the Mad Hatter in *Alice in Wonderland*? When the book was written, mercury was used in the manufacture of hats, making mercury poisoning an occupational hazard for millinery workers. Mercury is a well-established neurotoxin, yet many people are still unaware of its dangers. Mercury is emitted into the environment from coal-fired power plants, medical waste incinerators, and trash incinerators. Once it is released into the air, it falls with rain back into waterways (lakes, rivers, streams, oceans), where bacteria convert it into methyl mercury, a highly toxic form. Methyl mercury is then consumed by fish, which serve as an important source of mercury in the human diet. The problem is, methyl mercury lingers in our bodies and, over time, can damage nerve cells. Symptoms of methyl mercury poisoning include confusion, depression, fatigue, and memory loss. If it sounds a lot like symptoms of severe brain aging, in a sense it is. Mercury does its dirty work by promoting free radical production and inflammation; this is the same process that causes normal brain degeneration, but mercury does it much faster.

Many of us are walking around with mercury levels in excess of what's considered to be safe. A small study conducted by a physician in San Francisco found that 16 percent of those tested had mercury levels significantly higher than what the EPA deemed as safe (5 parts per billion [ppb]). If you eat fish every day, especially fish that is high in mercury, it is very easy to exceed this level. Having elevated levels of mercury doesn't mean that you're suffering from mercury poisoning, but it does mean that the toxic load on your body is higher than it should be and that you will not be able to properly defend yourself against free radicals. If you are constantly challenging your body with a high toxic load, your brain is likely to be spending all its time fighting free radicals as opposed to maintaining and repairing brain cells. As a result, whatever memory, concentration, or other cognitive problems you are having now will only get worse.

I know that fish is often touted as brain food because of its high omega 3 fatty acid content, but due to water pollution, many types of fish are contaminated with mercury. Some of the worst offenders include shark, swordfish, king mackerel, tilefish, halibut, and white albacore tuna (canned and fresh). Unfortunately, even though many of these fish are excellent sources of omega 3 fatty acids, I recommend that you don't eat them.

Fish that are lowest in mercury content include fresh, wild Pacific or Alaskan salmon, tilapia, and haddock. If you love tuna, stick to light tuna; it contains considerably less mercury than white tuna. Fortunately, these fish can be prepared in many different ways, so you won't get sick of them (or sick *from* them).

Mercury is a particular problem for young, developing nervous systems, and some scientists believe that it can trigger autism in children. The FDA recently warned pregnant women not to eat fish with the highest levels of mercury (shark, swordfish, halibut, king mackerel, and tilefish) because of potential harm to the fetus from mercury. According to a survey released by the Centers for Disease Control in Atlanta, 5 percent of all American women of childbearing age have mercury levels in their blood above the Environmental Protection Agency's safety threshold. Another 5 percent have mercury levels just below the thresh-

Fish with the Highest Mercury Content

Halibut	Swordfish
King mackerel	Tilefish
Shark	Tuna (white meat)

Moderate Mercury Content

Bass	Perch
Cod	Pike
Crab	Pollack
Flounder	Rainbow trout
Grouper	Scallops
Herring	Shrimps (canned or fresh)
Lobster	Snapper
Mahimahi	Sole
Orange roughy	Tuna (light—canned in oil or fresh)
Oysters (canned or fresh)	Turbot

Typically Low in Mercury

Sardines (canned in olive oil)	Tilapia
Alaskan sockeye salmon, Pacific salmon	Haddock

old. As many as three hundred thousand babies are born each year in the United States who are at risk of brain damage from mercury exposure in the womb. Although the FDA says it's safe for pregnant women to eat 12 ounces of any other kind of cooked fish, my advice is for them to stick to the fish with the absolute lowest levels of mercury.

Fish is not the only way mercury gets into our bodies. Your mouth may be filled with mercury. About 80 percent of all American adults have mercury amalgam (so-called silver) fillings in their teeth. How much of a threat does it pose to your nervous system? The American Dental Association has said that once mercury is sealed in an amalgam, it is locked in and cannot escape. Scientific studies have shown, however, that a significant amount of mercury vapor does escape from mercury fillings and is absorbed by the body. Chewing, consuming hot foods, even brushing your teeth causes mercury to be released from amalgam fillings. I personally am convinced that mercury amalgams don't belong in your mouth. More than 10 years ago, I had my mercury amalgam fillings removed and replaced with porcelain. Nevertheless, I don't recommend this process for everyone. First, unless it is done absolutely correctly by a dentist who specializes in mercury removal, you can end up being exposed to even more mercury than if you kept your fillings. The removal of mercury fillings can release a significant amount of mercury into your bloodstream. If you do have mercury removed, I recommend undergoing chelation therapy, a simple medical procedure in which an intravenous vitamin-mineral chelating solution is administered to rid your body of mercury residue. (Chelating agents are substances that bind to heavy metals and render them harmless.) Chelation therapy is performed as an outpatient procedure at a physician's office. It is painless but can take several hours. Basically, you simply sit in a chair as the vitamin-mineral is administered through an intravenous line connected to your arm. Chelation therapy has been used for several decades in the United States by alternative physicians as a treatment for heart disease. (See Appendix 3 on how to find a doctor who performs chelation therapy.)

Second, having your fillings removed can be expensive and not pleasant for people who dislike dental procedures. (If you want to find out more about having your mercury fillings removed, please see Appendix 3 for how to find a dentist who is properly trained in mercury-free dentistry.)

Watch the Aluminum

Aluminum is the most abundant metal on the planet. It is found naturally in food, soil, water, even the air we breathe. It is also in a wide variety of consumer products, from antacids to deodorants to processed cheeses to cookware. Like other metals, aluminum enhances the formation of damaging free radicals. Our bodies can handle a limited amount of aluminum and can excrete small amounts: daily ingestion of about 20 milligrams of aluminum poses no health risk. But modern-day consumption typically exceeds what many experts believe to be safe. For example, many brands of antacids contain aluminum; some brands contain as much as 200 milligrams of aluminum in a single tablet (see chapter 4 for a list of aluminum-containing drugs). If you routinely use antacids, you could be ingesting up to 4 grams of aluminum daily.

If you use a shampoo or a skin cream containing magnesium aluminum silicate or aluminum lauryl sulfate, or a deodorant containing aluminum chlorhydrate, or cook in an aluminum saucepan, you are ingesting more aluminum through your skin or food. For example, a study conducted at the University of Cincinnati Medical Center showed that if tomatoes are cooked in an aluminum container, you increase the aluminum content per serving by two to four milligrams. If there is aluminum in your water, you could be getting a few more milligrams daily. My point is, it all adds up.

What's wrong with consuming so much aluminum? Aluminum consumption may increase your risk for neurological disease. There is an abnormally high accumulation of aluminum in the brains of Alzheimer's patients, up to 30 times more than the normal level. There is a dispute in the scientific community as to whether the accumulation of aluminum in the brain is the cause of Alzheimer's or merely the result of the disease. It may be years before we know the real answer. Until we know for sure, however, I think it's smart medicine to assume the worst and reduce exposure to aluminum.

It's simple enough to avoid excess exposure to aluminum by reading labels and not using products that contain aluminum.

Antacids. If you use antacids frequently, that is, more than once or twice a month, switch to brands that contain calcium carbonate and avoid those that contain aluminum hydroxide. (Even better, try following a healthier diet that is less irritating to your gut. See the Better Brain meal plan in chapter 5.)

Water. To avoid excess aluminum, drink bottled. Even better, install a water purification system into your home that uses the reverse osmosis system. It eliminates most impurities from water, and the water tastes great. (See Appendix 3 for more information.)

Deodorants. There are several aluminum-free deodorants on the market, including those that contain pure baking soda. You can find several brands of aluminum-free deodorants at health food stores.

Shampoo. Some brands of shampoo, especially antidandruff shampoos, contain aluminum. Read the package labels carefully and avoid products that contain aluminum.

Cookware. I don't recommend using aluminum cookware because the aluminum can leach into your food. Instead, use stainless steel, copper, or glass cookware. If you use aluminum foil to store food, put a layer of wax paper in between to avoid aluminum coming in contact with your food.

Processed food. Aluminum is added as an emulsifying agent in many foods, including processed cheese (especially those that are single sliced). It is also found in many cake mixes, self-rising flour, prepared dough, nondairy creamers, pickles, and some brands of baking powder. Once again, your best defense is to read labels, and if a product contains aluminum, find a brand that doesn't.

For information on common household products that contain aluminum or other toxins, check out the Household Products Database at http:householdproducts.nlm.nih/gov/cgi-bin/household/brands?tbl=chem&id=2.

Lead Alert

Lead is a neurotoxin that is particularly harmful to children but can also affect adult mental performance. Exposure to lead can lower a child's IQ and cause learning and behavior problems. Until recently, lead has been widely used in paint and construction. Although there has been a conscientious effort on the part of government to raise awareness about the dangers of lead and to have it removed from apartment buildings and private homes, lead toxicity is still a problem. According to one study, between 1991 and 1994 about one million children between the ages of 1 and 5 had elevated blood lead levels. I've also seen my fair share of lead toxicity in adults. When I see patients suffering from memory problems or any brain function abnormality—especially those who live in an older home—I automatically test them for heavy metal toxicity. Not infrequently, we find that they have higher than normal levels of lead. The good news is, lead can be removed from the body through chelation (intravenous therapy), and once lead levels are normalized, many symptoms frequently disappear.

About 40 percent of the homes built before 1978 contain lead paint, and very often it has been painted over with other paint. Nevertheless, any lead residue can cause a problem. Paint can peel and chip off the walls and can be eaten by curious toddlers. Microscopic lead dust particles can be inhaled by both children and adults. There are easy at-home test kits that can help you determine whether your walls contain any lead paint residue. (See Appendix 3 for information.) If you find that you have lead paint in your home, it's best to get a professional to remove it from the walls so that lead dust particles don't spread throughout your house.

Although banned in new construction, lead or lead-lined pipes are still present in older homes, and the lead in them can leach into drinking water. Homes with plumbing that dates back before 1930 most likely have lead pipes. If possible, replace lead pipes with copper. *Beware:* copper pipes installed before 1988 may contain lead solder, which can leach into water.

If you're not going to replace lead pipes, at least don't use the tap water for drinking water. I recommend installing a water purification system in your home using the reverse osmosis method (see Appendix 3). In addition, never use hot water from the tap for drinking or cooking. Pipes exposed to hot water are more likely to leach lead into water than cold water pipes.

Ceramic dishes, especially those imported from other countries, may contain lead-contaminated glaze. Lead from the dish can leach into food or drink. These dishes are fine for display, but I don't recommend eating from them. If a ceramic dish does not specifically say it is lead free, don't use it.

MSG, Aspartame, and Hydrolyzed Vegetable Protein

Excitotoxins are chemicals that are added to food and that can cause brain cells to become overstimulated, triggering a surge in neurotransmitter activity that can harm healthy cells. Animal studies show that excitotoxins can damage the mitochondria, the energy-producing centers of brain cells, or neurons. There is some controversy as to whether consumption of excitotoxins is dangerous for humans. Since I consider these chemical additives to be completely unnecessary, I feel the best approach is to avoid them until we know for sure whether they are harmful.

There are three common excitotoxins in the food supply: (1) MSG (monosodium glutamate), a flavor enhancer; (2) hydrolyzed vegetable protein, which contains MSG and is used as both a flavoring agent and a filler in processed foods; and (3) aspartame, an artificial sweetener used in many diet products. MSG was originally used in Asian cuisines but has become a widely used seasoning in processed foods, from canned soups to frozen TV dinners. Aspartame is not an excitotoxin per se but can be metabolized into aspartate in the body, which is also an excitotoxin and is therefore included on this list.

Not everyone is sensitive to excitotoxins. Children with developing nervous systems may be more vulnerable to their negative effects, but so are some adults. For example, some people are highly sensitive to MSG. Immediately after eating a meal seasoned with MSG, these people may experience symptoms such as headache, dizziness, light-headedness, or heart palpitations. This phenomenon is so common that it is called "Chinese restaurant syndrome," so named because MSG is widely used in Chinese cooking. (I don't mean to single out Chinese restaurants. Many Chinese restaurants today no longer use MSG and say so on their menus, whereas food manufacturers are still adding MSG to their products.) Similarly, some people may get headaches or feel dizzy after drinking a beverage sweetened with aspartame, whereas others may not experience any symptoms at all. Whether or not you do experience symptoms, I think it's advisable to avoid using products with excitotoxins if you can, just as I think it's wise to avoid ingesting any unnecessary chemicals.

How do you know if a food contains MSG, hydrolyzed vegetable protein, or aspartame? If it is in a food product, it will be listed on the ingredients label. Many sugar-free, low-calorie foods (including diet soda) contain aspartame. Similarly, many processed foods, from soups to frozen TV dinners to seasonings, contain either MSG or hydrolyzed protein. The only way you can know for sure whether a product contains these ingredients is by reading the label. If you eat out at Asian restaurants, be sure to ask that your food be cooked without MSG.

Electromagnetic Fields: Is Your Cell Phone Frying Your Brain?

Television, computers, microwave ovens, cell phones, and hair dryers have all become the essentials of twenty-first-century life, and few of us are willing to give them up. The problem is, these modern conveniences emit radio frequency energy, or radio waves, which may be harmful to delicate brain cells and possibly increase the risk of brain tumors and other forms of cancer. Radio waves are a form of electromagnetic energy, which consists of waves of electric and magnetic energy moving through space. Humans have always been exposed to low levels of electromagnetic energy. Ultraviolet rays emitted from the sun are a source of electromagnetic energy, and the earth produces a weak electromagnetic field (EMF) as it rotates around its molten iron core. Since the dawn of the electricity age, around a century ago, our bodies have been exposed to much higher levels of radio waves than ever before. Power lines and electrical wires also produce electromagnetic fields. The reality is that you can't get away from them.

Even if you don't use any modern appliances, just living in the modern, industrialized world will expose you to your fair share of EMFs. Many scientists believe that EMFs are nothing to worry about. A major study conducted by the American Physical Society, which includes some of the nation's top physicists, reviewed more than a thousand scientific articles on the EMF-cancer connection and concluded that fears over EMFs and cancer were probably groundless. Nevertheless, there is an abundance of other research that shows that EMFs can alter cellular metabolism, which may be toxic to your brain. Electromagnetic fields enhance free radical production, which is the underlying cause of virtually all brain problems, from brain fog to memory loss to dementia. In a landmark study conducted at the University of Southern California School of Medicine, researchers found a substantial increase in Alzheimer's disease among people whose occupations exposed them to higher than normal levels of electromagnetic radiation, including electricians, machine operators, sewing machine operators, and welders. Some scientists speculate that EMFs could enhance the formation of beta amyloid, a protein that invades the brain's of Alzheimer's patients and enhances free radical production.

As a neurologist, I do worry about EMFs, particularly those emitted by cell phones. When you use a cell phone, you hold the phone right up against your head, and it directly bombards your brain with radio waves. Many of the 170 million Americans with cell phones spend hours a day on them. On any street or in any mall you will see people walking along with cell phones glued to their ears—especially teenagers. The question is: What is this doing to your brain?

Some studies have shown that the radio waves emitted by cell phones can cause cancerous mutations in animal cells, an indication that they may promote brain tumors in humans. To date, major epidemiological studies of cell phone users have not found an increase in the incidence of cancer of any kind, but none of these studies has investigated the impact of long-term exposure. Granted, the cell phones used today may be safer than the cell phones of yesteryear because they operate on much lower levels of power output, thereby probably reducing the risk of malignancy. And given the high number of cell phone users, we don't have an epidemic of brain tumors, at least not yet, although there is certainly an increase in the incidence of brain tumors. Nevertheless, living tissue is very sensitive to electromagnetic energy, and brain cells in particular are very delicate. I am particularly concerned about the widespread use of cell phones among children, whose brains and nervous systems are not yet fully formed. Even if cell phones don't cause brain tumors, I believe that scientific evidence clearly demonstrates that these phones can certainly accelerate the growth of existing tumors. About 10 years ago, when cell phones had just become popular, I treated two patients with malignant brain tumors that exactly corresponded to the position of the base of the antenna of their handheld phones. I was so troubled by what I had seen that I testified as an expert witness in a class action suit brought against several of the large manufacturers of cellular phones.

On their website, the FDA offers a frank analysis of cell phone safety data. Although the FDA website offers data supporting the safety of cell phones, under the question "Do wireless phones pose a health hazard?" the FDA offers this cautious answer. "The available scientific evidence does not show that any health problems are associated with using wireless phones. There is no proof, however, that wireless phones are absolutely safe." To me, this is not exactly a ringing endorsement.

I now use a cell phone myself, but I try to use it safely. Here's some advice on how to enjoy the convenience of a cell phone without exposing yourself to too much risk.

Don't use your cell phone as your primary phone. Cell phones are not only a great convenience but have become very inexpensive. You can literally talk for thousands of minutes a month for pennies a conversation. However, I would not allow myself to be lured in by cheap prices. Limit conversations on your cell phone to no more than a few minutes a day. Use the phone only when you have to, and don't use it for lengthy conversations.

Use an earphone. Radio waves are created as a result of the movement of electrical charges in antennas. As the waves are created, they radiate away from the antenna. The closer you are to the antenna, the greater your exposure. The same innovation that makes cell phones portable— their built-in antennas—could also pose a health hazard. When you use your cell phone, attach an earplug to the phone so you don't have to hold it right next to your head. *Don't clip the phone on your clothes next to your body*—it simply exposes a different part of your body to radio waves. Put the phone on a table or a cradle away from your body.

Beware of phone shields Some companies market so-called shields or special cases to place around your phone to reduce radio wave absorption. The FDA warns that these shields may interfere with the operation of the phone, forcing the phone to increase boost power to compensate, leading to greater radio wave exposure.

What About Other Sources of EMFs?

If you watch TV or use a computer or a hair dryer, you are exposed to low levels of EMFs. We can't turn back the clock to the days before electricity, and I don't think most of us would want to, but do exercise caution when using EMF-emitting appliances.

Make your bed a safe zone. Reduce your exposure to EMFs while you sleep. Don't sleep with a clock radio or telephone answering machine right next to your head. Move your clock or answering machine to the foot of your bed, or at least three feet away from your head. I don't recommend sleeping with an electric blanket because you are directly exposing yourself to an electrical field for an extended period of time. If you love your electric blanket, use it for a few minutes before you go to sleep to warm up your bed and then unplug it for the night.

TV tips. Don't sit too close to the television. Maintain a distance of about 6 feet between you and the screen.

A safer computer. If you routinely use a computer, use one with a flat screen, which emits far lower levels of EMFs than the old-style screens with a cathode ray tube (CRT). They are somewhat more expensive than the CRT screens but are coming down in price; they are well worth the extra dollars in terms of safety. Whatever type of screen you use, be sure to sit at least 3 feet away.

Hair dryer. If you use a hair dryer, keeping it on the lowest setting will reduce EMF exposure. If you let your hair air dry for a few minutes before using the blow dryer, you will not need to use the hair dryer for as long a period of time.

What about the EMF exposure that you can't control? I personally wouldn't want to live or work in a building near high-tension electric wires or transformers, but if you do, you should be especially careful about exposure to EMFs from other sources, such as a cell phone. As far as I'm concerned, taking these simple precautions is just good, common-sense "preventive medicine."

The Brain Workout

YOU MAY GO to the gym or to an exercise class to work out your muscles so that they don't atrophy with age, but what have you done for your brain lately? If you are not doing some form of brain fitness, you run the risk of losing brain power with each passing year. Just as your muscles go flabby without exercise, so do your brain cells. If you are having difficulty remembering things, if you aren't as focused or alert as you used to be, or if you feel that your creative juices are drying up, it's a sign that your brain cells need some toning up.

A good mental workout will give your brain cells a power boost. When your brain is stimulated by a new challenge, you make new branches from your neurons, called dendrites, that are essential for processing and assimilating information. Think of it as laying down new telephone wire so that your brain cells can communicate better or, in modern-day terms, adding RAM to your hard drive. It's easier to make dendrites when you are young, before the brain has been worn down by free radical attack and a slowdown in energy production. In fact, at one time it was believed that there was a point in life when you stopped making dendrites. Although it becomes more difficult, we now know that you can make dendrites—and even new neurons—well into old age if you keep your brain active. *You need to keep making new dendrites to perform well right now and to stave off brain aging later.* Scientists theorize that if you go into old age armed with dendrites to spare, you will remain mentally resilient and be less likely to develop dementia.

How do you make new dendrites? Performing mentally stimulating activities can prevent the age-related decline in dendrite production and reinvigorate an aging brain. It's never too late to grow new dendrites, but to achieve the best results, you need to engage in the right kind of mental activity.

So what should you do to keep your brain fit? I have created a regimen of four brain-boosting exercises for my patients to accomplish specific goals. Exercises 1 and 2 are designed specifically to enhance memory. Exercises 3 and 4 are designed to improve mental speed and accuracy. All it takes is 15 minutes a day, or three five-minute "brain breaks," to do your brain workout. If you do both sets of exercises regularly, not only will you see a definite improvement in mental function but you'll protect your brain against getting "flabby" and out of shape.

Two Memory-Boosting Exercises

Although both exercises will help improve overall memory and sharpen your ability to retrieve information on demand, exercise 1 is designed to boost your ability to recall numbers, while exercise 2 is geared to help people remember names better.

I don't recommend that you do exercises 1 and 2 on the same day. If you do, it will only confuse you and slow down your progress. You can alternate exercises daily (for example, do exercise 1 on odd days and exercise 2 on even days) or alternate between them weekly, or even monthly. If you have a specific problem—for example, if you have more difficulty remembering numbers than names—you may choose to master exercise 1 before starting exercise 2.

These exercises will only work if you do them consistently. Set aside specific times each day to do your brain workout. I recommend that you do your brain exercises immediately after breakfast, lunch, and dinner. You will be nourished, relaxed, and alert and able to perform at your best.

Here's what you need:

> *Exercise 1:* Regular deck of playing cards (remove aces and picture cards)
>
> *Exercise 2:* Full deck of playing cards and a phone book

Exercise 1: Improve Your Ability to Recall Numbers

For exercise 1, *remove all aces and picture cards from your deck of cards.* This exercise should markedly improve your ability to remember numbers, street addresses, appointments, and other vital information.

Step 1: Choose a card. Each morning, following a brain-boosting breakfast (see chapter 5) and your morning nutritional supplements (see chapter 6), take out your deck of cards. Randomly select one card and remove it from the deck.

Step 2: Write down the number. Look at the number on the card and write it down on a piece of paper. Pay no attention to the suit. Your goal is to be able to recall the number at the end of the day.

Step 3: Be sure to say the number out loud after you write it down. Speech activates different areas of the brain and will help enhance memory.

Step 4: Make a mental file. To enhance your ability to recall this number, you will have to mentally process this information in a way that will help you retrieve it later. In other words, you have to store it in a safe place where you can find it when you need it. You do this all the time in everyday life. For example, when you have a piece of paper on your desk that you don't want to misplace or a document on your computer that you want to save, you create a file for it. You probably keep related documents in the same file. The reason people use files is so they can know where they need to go to access specific information. In much the same way, I want you to create a special place in your brain for the number you have just selected. It is helpful to actually visualize a file folder and to name it. For the purposes of this exercise, name it the "number file." Close your eyes and create a mental image of the number you selected from the deck of cards being placed in the file.

Creating the mental image of the number file will profoundly improve your recall ability.

Step 5: Recall your number. Your number file now contains *one* number. Immediately after lunch, recall the number. The key to success is not to first think of the number but to ask yourself where is the number stored? The number is stored in the number file. When the image of the file pops into your head, you will see the image of the number. Recall the number once again in the evening after dinner. Remember to think first of the image of the file and then mentally retrieve the number.

Step 6: Your daily workout. Repeat this exercise daily, choosing a different card from the deck each time. As with any new exercise, you may find this one challenging at first, but it will soon become easier. Do the exercise using a single number until you are able to remember each day's number for at least six out of seven days.

Step 7: Increasing the workout. When you can remember the number of the one card for six out of seven days, begin drawing *two* cards from the deck each morning and work on remembering the two digits, in the order that you picked them. Again, pay no attention to the suit. This time, you are filing two numbers in your number file. Follow steps 2, 3, and 4. Remember to think of the image of the file first before trying to recall the numbers. Again, continue the exercise on a daily basis until you are able to recall the two numbers after lunch and after dinner for six out of seven days. When you can recall two numbers for six out of seven days, add a *third* number to your exercise regimen. When you are able to correctly remember the sequence of three numbers for six out of seven consecutive days, you are ready to move on to a new challenge.

Step 8: A new challenge. When you have successfully learned three numbers for six out of the seven days, you are ready to add a fourth number. Now that you have more numbers to remember, you need to set up a more efficient filing system. Create an image in your mind of the first three numbers going into the file as a single group. Then visualize the fourth number going into the file separately. Once you are able to correctly remember four numbers for six out of seven days, continue adding additional numbers up to a maximum of seven. The numbers should be "filed" and remembered as two groups, with the first group containing the first three numbers and the second group containing the remaining numbers. Again, when it's time to recall the number sequence, think first of the image of the file where the numbers are stored. Then think of the first group of numbers being filed, and then think of the second group of numbers being filed.

Step 9: Keep going. When you have mastered the ability to memorize a new group of seven numbers each day, you should move on to my other brain-building exercises; but continue to do this exercise using seven numbers at least once a week. While this exercise may seem simple, its implications are profound. One obvious benefit is that learning to recall seven numbers in proper sequence will help you remember phone numbers and, in particular, will help you to retrieve the *right* number when you need it. (This is a particularly useful exercise for those of you who frequently think that you are dialing one phone number when you are actually dialing another.) The act of creating a "file" will dramatically improve your recall ability, and not just for numbers but for other facts as well. For example, if you frequently misplace your reading glasses, use the file system to help you keep track of them. Set up a mental file for "reading glasses." Every time you put your reading glasses down, create a mental image of where you are placing your glasses, and file that image away into your reading glasses file. When it's time to locate your glasses, visualize your reading glasses file in your head, and you will be able to retrieve the location of your glasses.

Even after just one week of practicing this exercise you can begin to use this technique to remember all sorts of information that may be important to you. You will be amazed at how easy it is to recall facts and figures when you begin using this powerful mental device to both store and retrieve information.

Exercise 2: Placing the Name with the Face

This brain exercise is great for people who are constantly saying, "I can place the face, but not the name!" This exercise will teach you a technique that will help you to remember names of people immediately after you are introduced to them and to better retain those names for extended periods of time. This exercise should take no more than 10 minutes a day.

For this mental exercise, you will need a deck of cards and a phone book. You don't need to remove any cards from the deck.

Step 1: Choose the name and card. Each morning after your brain-boosting breakfast (see chapter 5) and your morning nutritional supplements, open the phone book and randomly select a name. Then choose a card at random from the deck.

Step 2: Write down the name and card. On a piece of paper, write down just the *first* name of the person that you selected from the phone book, and only the *suit* of the chosen card on a piece of paper. For example, if you selected the name Mary Smith and picked the ten of hearts from the deck of cards, you will write down "Mary Hearts."

Step 3: Be sure to say the name and suit out loud after you write it down. Speech activates different areas of the brain and will help enhance memory.

Step 4: Make a mental file. Once again, you will set up a mental file for "names." Create a mental image of both the name and the suit being placed into your names file.

Step 5: Recall the name. After lunch, try to remember both the name and the suit. Begin by creating a picture in your head of the names file, and imagine yourself opening the file. Repeat the exercise in the evening after dinner. If you miss the recall at lunch but get it in the evening, or vice versa, it counts as a successful recall.

Step 6: Your daily workout. Do this exercise every day. Once you are able to remember the name and the suit correctly on six out of seven days, you are ready for a new challenge.

Step 7: A new challenge. This time, select the *entire* name, first and last, from the phone book and pick one card from the deck. Write down the first and last name of the person you selected and the suit of cards. For example, if you selected Mary Smith and the ten of diamonds, you would write down "Mary Smith" and "diamonds." In your mental names file, visualize the name Mary Smith and the image of diamonds being placed in the file. After lunch, try to recall the name and the suit of the card. Begin by creating a picture in your head of your names file, and mentally opening the file. Mary Smith with the diamonds face should pop out of the file. Over time this exercise will develop your ability to visualize an object (the image of spades, hearts, diamonds, or clubs) along with a complete name. It is this type of memory that will readily allow you to remember someone's name and connect it to the image of his or her face.

Using This Exercise in Real Life

Once you have mastered this exercise, you should be able to easily connect faces and names. When you are introduced to a new person, the key to success is to visualize the name being put into the special file, followed by the image of the person's face. Do this the moment you meet someone. Say their name out loud before mentally putting it in the name file. This simple exercise will very quickly help you recall a person's name when seeing his or her face, as well as retrieve the mental image of someone's face when you are given someone's name.

Exercises to Speed Up Reaction Time

If you feel that you are slowing down, that is, if performing simple tasks like filling out your expense sheet at work or a form at the bank takes you longer than it used to, or if you are easily distracted (that is, you can't stay on task if you're interrupted), exercises 3 and 4 are just what the doctor ordered. These two exercises will significantly enhance the speed and accuracy of your brain's performance.

For exercises 3 and 4 you need a stopwatch and a full deck of playing cards (don't remove the picture cards or aces). You can get an inexpensive stopwatch from most sporting goods or electronics stores. Many common electronic watches also have a stopwatch function.

Exercise 3

Your task is to separate the deck of cards into the four suits in as short a time as possible. That is, you will be making a separate pile each for diamonds, spades, clubs, and hearts.

Step 1. Put your stopwatch in a handy place near the deck of cards. You will be starting and stopping the watch yourself.

Step 2. Begin with the deck facing you. Make sure it is well shuffled.

Step 3. When you are ready, start the stopwatch and begin separating the deck into four separate piles for each of the suits (diamonds, spades, clubs, and hearts).

Step 4. Work as quickly as you can. When finished, stop the watch and record the time. Part of the test includes the time it takes for you to start and stop the stopwatch, so don't have someone do that for you.

Step 5. Do this exercise three times daily. (I recommend that you do it after you do your memory exercises.)

Track your progress. This simple exercise will greatly enhance your ability to process information and make decisions quickly. There are no specific "norms" for this test, but a typical young adult can easily complete this exercise cold in 35 seconds, and I see no reason why any adults shouldn't strive to do the same. Don't be discouraged if it takes you a minute or more when you first begin. The more you do it, the more you will improve, and the faster you accomplish the task, the more enhanced will be your accuracy and the speed of your computer brain.

Keep doing this exercise indefinitely. This simple exercise (as well as exercise 4) should be continued indefinitely to keep your brain at peak performance.

Exercise 4

A variation on exercise 3 with an interesting new twist. Instead of just sorting the deck by suit, you will separate the deck into four separate piles. All pictures will be in one pile, all aces in a second pile, all odd-numbered cards in a third pile, and all even-numbered cards in a fourth pile.

Step 1. Put your stopwatch in a handy place near the deck of cards. You will be starting and stopping the watch yourself.

Step 2. Begin with the deck facing you. Make sure it is well shuffled.

Step 3. When you are ready, start the stopwatch and begin separating the deck into four separate piles divided by picture cards, aces, odd-numbered cards, and even-numbered cards.

Step 4. Work as quickly as you can. When finished, stop the watch and record the time. Part of the test includes the time it takes for you to start and stop the stopwatch, so don't have someone do that for you.

Step 5. Do this exercise three times daily. (I recommend that you do it after you do your memory exercises.)

This exercise is similar to exercise 3 but somewhat more challenging since it draws on and enhances the function of slightly different areas of the brain. If you have the time, it would be ideal to do both of these exercises together, right after your memory exercises, or you can alternate between exercise 3 and exercise 4 every day. I can't tell you what is considered "normal" for this test; it will vary depending on a variety of factors that have an impact on brain function like age, medications, supplements, and so on, but I will tell you that a typical sharp young adult can complete this exercise cold in about 27–30 seconds, and I have seen several sharp-minded older adults achieve the same speed after a bit of practice.

The more proficient you become at this exercise, the more you will see an improvement in brain power in many different areas. You'll find it much easier to stay on task, especially when faced with time constraints. Distractions that in the past may have interrupted your concentration will not bother you as much. Most important, these exercises will allow you to carry out several tasks simultaneously with far less effort than in the past. You'll find yourself being able to perform mental calculations in the middle of a conversation, write notes to yourself while speaking on the phone, and consider many more options before making a quick decision.

More Brain Workout

What do I do to keep my brain fit? In addition to the card games described here I use a computer-based mental training program called BrainBuilder 3.0, a software program for home computers. BrainBuilder provides several mind-challenging programs that are actually fun to do. It's not just a game; these mental tests are designed to help you shorten your reaction time as well as enhance short-term memory. I recommend that you try to incorporate BrainBuilder, or a similar program, into your brain workout. The more variety you can add to your brain workout, the more likely it is that you'll stick to it, and the better your result. (See Appendix 3 to find out how you can purchase BrainBuilder and similar programs.) In addition, online brain training and testing is available at www.TestYourBrain.com. This unique web site will allow you to assess your brain function, participate in brain training exercises, and follow your progress.

Four Medical Tests That Can Save Your Brain

HOW FAST and how well is your brain aging? Is it winning or losing the war against free radicals? Are your brain—and your body—being slowly destroyed by inflammation? Do you carry a gene that may make you vulnerable to Alzheimer's and other neurological diseases because you do not produce enough protective antioxidants? There are four simple, safe, and noninvasive medical tests that can provide answers to these questions and reveal vital information about the health of your brain and your susceptibility to brain aging. I urge all my patients to have these tests, and I feel that they should be part of every routine physical examination.

Each one of these tests can detect potential brain and neurological problems long before you may experience any symptoms. Armed with this information, you can take the necessary steps to fix these problems early on, before they can cause significant damage. In some cases, these tests should be repeated to monitor whether or not your program is working, and whether or not you need to consider taking stronger measures. To me, this is true preventive medicine, and how I believe medicine should be practiced in the twenty-first century.

If you've taken the Brain Audit (chapter 2), you have already identified your tier of the Better Brain program. *Your results on these medical tests may put you at an increased level of risk not detected on the questionnaire. Depending on your results on the medical tests, you may need to follow a different tier. I will tell you exactly how to interpret your test results so that you know exactly which tier is the right one for you.*

Many insurance companies cover some or all of these tests. Check with your insurance provider to see which tests are covered under your plan.

The Lipid Peroxide Test: Are You Under Free Radical Attack?

When your brain is stuck in low gear, when you're forgetting things from minute to minute and stumbling over your words, it's a sign that your brain is overrun by free radicals. How do you know for sure? The lipid peroxide test measures the degree of free radical activity in your brain. Why do you need to know this? Free radical damage is not only affecting your day-to-day brain function but has been linked to many serious ailments, from premature brain aging to Alzheimer's disease, Parkinson's disease, and even heart disease. The lipid peroxide test can also tell you whether you are taking enough antioxidants to stop the free radical damage.

Lipid peroxide, a chemical that is excreted in urine, is an accurate marker of free radical activity in the fatty tissue of your body. Since your brain is primarily made of fat, the test is an excellent way of determining free radical activity in your brain. The higher the level of lipid peroxide in urine, the higher the degree of free radical assault on the brain and nervous system. It could take years or even decades before you see the end results of this free radical activity in terms of symptoms. The lipid peroxide test provides an early warning system, alerting you to the degree to which your brain might be slowly and insidiously damaged by free radicals. The lipid peroxide test is also a measure, albeit indirect, of the strength of your antioxidant defense system. If you are getting adequate amounts of antioxidants in food and supplements, your body should be able to control free radicals. There are other factors that could be causing you to produce excess free radicals. Drugs—prescription and over the counter—can lower your body's natural antioxidant defenses (see chapter 4). Illness and stress can also promote free radical production and sap you of vital antioxidants. If you have high levels of lipid peroxide, it's a sign that you need to change your diet, boost your intake of antioxidant supplements, and take inventory of your lifestyle to see if you are being exposed to toxins (see chapter 8) or are under undue stress (chapter 7).

In a groundbreaking study published in the *Archives of Neurology* (June 2002) researchers found a direct correlation between high levels of lipid peroxide and mild cognitive impairment (MCI) in elderly patients, a condition now recognized as a precursor to Alzheimer's disease. About half of all people with this problem will go on to develop Alzheimer's. More important, the researchers noted that the only discernible difference between elderly people with MCI and their peers who did not have cognitive problems "was . . . this marker of oxidative stress." They go on to say that "this observation would suggest that brain oxidative damage is an early event in the development of AD [Alzheimer's disease]-like pathology." In other words, this test could be a way of detecting Alzheimer's in its earliest stages when it still may be possible to stop or slow down the damage by bolstering the body's antioxidant defenses. This is preventive medicine at its finest.

Which Test?

There are three types of lipid peroxide tests. The first is performed at the doctor's office, the second is performed at home but sent to a laboratory for analysis, and the third is entirely performed at home and is similar to a home pregnancy test kit. I'll review them all here.

Do not take any multivitamin or antioxidant supplements up to 24 hours before taking any of these tests because it could skew your results.

At the Doctor's Office

I perform a lipid peroxide test on virtually all of my patients. I use the Oxidative Stress Test distributed by Great Smokies Diagnostic Laboratory. As part of the routine physical, patients submit a urine sample that is then sent to Great Smokies to be analyzed. The results are based on a numerical rating.

> *NO RISK:* a score of 0–3 indicates a very low level of free radical activity.
>
> *LOW RISK:* a score of 4–6 indicates a low level of free radical activity.
>
> *MODERATE RISK:* a score of 7–9 indicates a moderate level of free radical activity.
>
> *HIGH RISK:* a score of 9 or over indicates a high level of free radical activity.

Which tier of the Better Brain program is right for you? When you get your test results back from your physician, your score on the lipid peroxide test may affect which tier of the Better Brain Program you should follow.

No risk: 0–3. If your lipid peroxide test indicates free radical activity in the optimum or no-risk range, stick to the tier determined by your score on the Brain Audit. This result does not affect your level of risk.

Low risk: 4–6. If your lipid peroxide test indicates free radical activity in the low-risk range, stick to the tier determined by your score on the Brain Audit. This result does not affect your level of risk.

Moderate risk: 7–9. Increases your risk by one tier. If your result on the lipid peroxide test indicates free radical activity in the moderate-risk range, it increases your risk level by one tier. If, based on the Brain Audit, you are now following Tier 1, you will now follow Tier 2. If, based on the Brain Audit, you are now following Tier 2, you will now follow Tier 3.

High risk: 9 plus. Follow Tier 3. If your result on the lipid peroxide test indicates free radical activity in the high-risk range, *follow the Tier 3 program for people at higher risk regardless of your score on the Brain Audit.*

At Home–Laboratory Analyzed Test

Great Smokies offers an at-home version of the same oxidative stress test I use in my office: the AntiOxidant Check from their Body Balance division. This is a top-notch test that you can purchase online (www.ULPTEST.com) or at health food stores and pharmacies. The kit includes a cup for a urine sample. Once the sample is taken, you seal it and return it in a prepaid mailer back to Body Balance laboratories for analysis. Within a week or two you receive your results, which are based on a numerical rating.

NO RISK: a score of 0–3 indicates a very low level of free radical activity.

LOW RISK: a score of 4–6 indicates a low level of free radical activity.

MODERATE RISK: a score of 7–9 indicates a moderate level of free radical activity.

HIGH RISK: a score of 9 or over indicates a high level of free radical activity.

Which tier is right for you? Your score on the lipid peroxide test may affect which tier of the Better Brain Program you should follow.

No risk: 0–3. If your lipid peroxide test indicates free radical activity in the optimum or no-risk range, stick to the tier determined by your score on the Brain Audit. This result does not affect your level of risk.

Low risk: 4–6. If your lipid peroxide test indicates free radical activity in the low-risk range of free radical activity, stick to the tier determined by your score on the Brain Audit. This result does not affect your level of risk.

Moderate risk: 7–9. Increases your risk by one tier. If your result on the lipid peroxide test indicates free radical activity in the moderate-risk range, it increases your risk level by one tier. If, based on the Brain Audit, you are now following Tier 1, you will now follow Tier 2. If, based on the Brain Audit, you are now following Tier 2, you will now follow Tier 3.

High risk: 9 plus. Follow Tier 3. If your result on the lipid peroxide test indicates free radical activity in the high-risk range, *follow the Tier 3 program for people at higher risk regardless of your score on the Brain Audit.*

Complete At-Home Test Kits

As of this writing, there are two different types of lipid peroxide tests that can be performed and analyzed entirely at home: The Vespro Free Radical Test and the OxyStress Test marketed by North American Pharmacal. Both can be purchased by mail, online, or at some natural food stores.

The Vespro Free Radical Test contains a small plastic cup and a glass dropper tube with an attached ampule. You fill the cup halfway with urine (from the first morning urine of the day) and then place a dropperful of urine in the ampule. After five minutes, the ampule turns different colors, from white to dark pink, depending on the degree of oxidation. The Vespro test does not give a numerical grade. Instead, it rates your level of risk as follows.

OPTIMUM. (The ampule remains white.) Optimum free radical activity puts you at NO RISK.

LOW. (Ampule turns very light pink.) Low free radical activity puts you at LOW RISK.

MEDIUM. (Ampule turns deeper shade of pink.) Medium free radical activity range puts you at MODERATE RISK.

HIGH. (Ampule turns fuchsia or intense shade of pink.) High free radical activity puts you at HIGH RISK.

Which tier of the Better Brain program is right for you? Your score on the Vespro lipid peroxide test may affect which tier of the Better Brain Program that you should follow.

OPTIMUM LEVELS. If your lipid peroxide test indicates free radical activity in the optimum or no risk range, follow the tier based on your score on the Brain Audit. This result does not affect your level of risk.

LOW LEVELS. If your lipid peroxide test indicates free radical activity in the low range, follow the tier based on your score on the Brain Audit. This result does not affect your level of risk.

MEDIUM LEVELS—Increases your risk by one tier. If your result on the lipid peroxide test indicates free radical activity in the moderate-risk range, it increases your risk level by one tier. If, based on the Brain Audit, you are now following Tier 1, you will now follow Tier 2. If, based on the Brain Audit, you are now following Tier 2, you will now follow Tier 3.

HIGH LEVELS—Follow Tier 3. If your result on the lipid peroxide test indicates free radical activity in the high-risk range, *follow the Tier 3 program for people at higher risk regardless of your score on the Brain Audit.*

The OxyStress test, produced by North American Pharmacal, also contains a plastic cup, a dropper tube, and an ampule, but it is plastic, not glass.

NO RISK (the ampule is white): a score of +0 or least oxidation means that no free radical activity was detected.

LOW RISK (the ampule turns very light pink): a score of +1, or low oxidation, indicates low free radical activity.

MODERATE RISK (the ampule turns a deeper pink): a score of +2, or moderate oxidation, puts you in the medium free radical range.

HIGH RISK (the ampule turns fuchsia, the deepest shade of pink): A score of +3 or high oxidation puts you in the high free radical range.

Which Tier of the Better Brain Program Is Right for You?

Your score on the lipid peroxide test may affect which tier of the Better Brain Program you should follow.

If you are in the *optimum, no-risk* range, this result does not affect your level of risk. Follow the tier determined by your score on the Brain Audit.

If you are at *low risk,* this result does not affect your level of risk. Follow the tier determined by your score on the Brain Audit.

If your result is in the *medium-* or *moderate-risk* range, it increases your risk level by one tier. If, based on the Brain Audit, you are now following Tier 1, you will now follow Tier 2. If, based on the Brain Audit, you are now following Tier 2, you will now follow Tier 3.

If your result indicates free radical activity in the *high-risk* range, *follow the Tier 3 program for people at higher risk regardless of your score on the Brain Audit.*

Follow-up

If your results are in the no-risk or low-risk range, you should retake the lipid peroxide test every year. If your results are within the moderate- to high-risk range, retake the test every three months until your levels have been normalized. The overwhelming majority of people will respond fairly quickly to the antioxidant supplements in Tier 2 or Tier 3, depending on which program they are following.

If you are in the high-risk group and are following the Tier 3 supplement regimen and your lipid peroxide levels are not normalized within three months, follow the Tier 3 Plus program shown here. Follow this enhanced regimen for another four months and then retake the lipid peroxide test. If your lipid peroxide level is not reduced at that

Tier 3 Plus		
SUPPLEMENT	A.M.	P.M.
DHA	300 mg	300 mg
Co-Q10	200 mg	200 mg
Vitamin E	400 IU	400 IU
Vitamin C	500 mg	500 mg
Alpha lipoic acid	200 mg	200 mg
N-acetyl-cysteine (NAC)	800 mg	800 mg
Acetyl-L-carnitine	400 mg	400 mg
Phosphatidylserine	100 mg	100 mg
Ginkgo biloba	60 mg	
Vitamin D	400 IU	
Vinpocetine*	5 mg	5 mg

Vitamin B-complex supplement:**		
B1 (thiamine)	50 mg	
B3 (niacin as niacinamide)	50 mg	
B6 (pyridoxine)	50 mg	
Folic acid	400 mcg	400 mcg**
B12 (cobalamin)	500 mcg	500 mcg***

*Vinpocetine is required only for people with high homocysteine or a history of vascular dementia or coronary artery disease. It should be avoided by people taking Coumadin, a blood thinner.

**Look for a B-complex supplement containing several B vitamins in one capsule or pill.

***In addition to your B complex in the A.M., take an additional 500 mcg of B12 and 400 mcg of folic acid in the P.M.

point, I recommend that you consider getting intravenous glutathione, which can be done in your doctor's office or at home. The protocol is described in Appendix 4.

Homocysteine Check: Are You Getting Enough B Vitamins?

If you've had a physical examination within the past five years, and your doctor hasn't checked your homocysteine level, he (or she) should be ashamed. I know that these are strong words, but given everything we know about the relationship between elevated homocysteine levels and the health of your brain and your body, it's absolutely inexcusable that this test is not a routine part of every annual physical. Could it be that this test is overlooked because the solution to high homocysteine is a simple, inexpensive vitamin regimen that is not being promoted by a pharmaceutical company with a hefty advertising budget? I hate to sound cynical, but once you learn a bit more about homocysteine, you'll understand why I feel that this test is so important, and why I am so angry that so few doctors offer it to their patients.

Homocysteine is an amino acid that is produced by every cell in the body as a normal part of the methylation cycle. *Methylation* is a term for a basic yet critical chemical reaction in the body, involving the passage of a methyl group—one carbon and three hydrogen atoms—from one molecule to the other. Methylation is vital for life and is involved in hundreds of different processes in the body, from producing neurotransmitters in the brain to preserving bone health and turning on the genes that help repair DNA. In the body, folic acid and vitamin B12 break down homocysteine into another amino acid, methionine, which is the building block of S-adenosyl-methionine (SAM-e), an amino acid associated with mood. SAM-e, in turn, increases the activity of an enzyme that converts methionine into beneficial glutathione. As long as there are ample B vitamins on hand, the body has a way of taking care of excess homocysteine.

What happens if there are not enough B vitamins available to break down homocysteine? Elevated levels of homocysteine can shrink your brain, dull your reflexes (especially those involving hand-to-eye coordination), and lead to depression. It can also *double* your chances of developing Alzheimer's disease. Simply put, in high amounts, homocysteine can kill you. It becomes toxic to the endothelial cells that line your arteries, and this promotes atherosclerosis, or hardening of the arteries in your brain and heart. Excess homocysteine can enhance free radical damage within the artery, which increases the formation of plaque and triggers an inflammatory response by the immune system. Too much homocysteine and/or too few B vitamins are associated with an increased risk of stroke, heart disease, depression, and even certain types of cancer, as has been documented in countless medical journal articles.

Unfortunately, many people have high homocysteine levels and don't even know it, so they don't take any steps to correct it. Typically, they are deficient in B vitamins. Our modern, highly processed diet is often lacking in B vitamins, such as folic acid, which are present in whole grains, green leafy vegetables, and dried beans. Red meat is a good source of B12 (lean cuts only, please!). But as people age, they may have difficulty absorbing this vitamin from food because their stomachs produce less hydrochloric acid, which is necessary for the complete breakdown of proteins. This condition, known as hypochloridemia, can cause the same symptoms of indigestion characterized by *excess* stomach acid, and many people erroneously treat it by taking over-the-counter antacids. This only makes matter worse because it further depletes acid production and interferes with the absorption of B12. Moreover, many commonly used prescription and over-the-counter drugs that you may very well be taking deplete B vitamins and thereby elevate homocysteine (see the list of drugs that deplete vitamin B in chapter 4). It has long been known that people suffering from Parkinson's disease generally have higher-than-normal levels of homocysteine, which is probably why Parkinson's patients have a much greater risk of stroke. We now know that the mainstay treatment for Parkinson's disease, the drug L-dopa, actually *raises* homocysteine, further jeopardizing the health of these patients. This is why I believe that checking homocysteine should be mandatory in the care of Parkinson's disease.

The good news is that elevated homocysteine levels can easily be lowered by taking the right combination of B vitamins and eating a diet rich in B vitamins. It's that simple.

What is normal? Homocysteine levels are measured in micromoles per liter of blood. The test is simple. As for a cholesterol test, your doctor takes a blood sample from you and sends it to a laboratory for analysis. A homocysteine level of over 9 micromoles per liter dramatically increases your risk of neurological problems and should therefore be treated. Anything under 9 is optimal. Please note that some laboratories consider 11 micromoles per liter to be within normal range; however, the latest studies suggest that they are wrong and that this is too high.

Which Tier of the Better Brain Program Is Right for You?

Your test result may affect which tier you should follow. If your homocysteine level is over 9, you must take immediate steps to lower it. *An elevated level of homocysteine increases your risk level by one tier.*

If you have a 9 or higher homocysteine level and your results on the Brain Audit put your risk level at Tier 1, move up to Tier 2. If you have a 9 or higher homocysteine level and your results on the Brain Audit put your risk level at Tier 2, move up to Tier 3.

Follow-up

If your homocysteine is normal, redo the test annually when you have your yearly physical. If you begin taking a drug that can raise homocysteine levels, retake the test sooner, about three to four months after starting the drug.

If you have high homocysteine, redo the test in two months after following the Tier 2 or Tier 3 program. If your homocysteine levels are normal, continue with your program.

If your homocysteine levels are still too high, add the following supplements to your daily regimen.

> *Folic acid:* add an *additional* 800 mcg for a total of 1,200 mg daily.
> *B6:* add an *additional* 50 mg for a total of 100 mg daily.
> *TMG:* 500 mg daily

Trimethylglycine (TMG) is a naturally occurring substance found in plants and animals. It can convert harmful homocysteine back into methionine, which is beneficial to the body.

Recheck your homocysteine levels in two months to make sure that your homocysteine has been brought down to 9 or under. If your homocysteine is normal, you can discontinue taking the additional supplements, but do check it again in two months.

If your homocysteine is still too high, you may need weekly vitamin B12 shots until your homocysteine is normal. After your homocysteine is normalized, you can resume taking the standard B-complex supplement, but you may still need monthly vitamin B12 shots.

The ApoE4 GENOTYPE:
The "Alzheimer's Gene"

ApoE is a protein found in the body and it is involved in the transport of cholesterol. There are three different types of the ApoE proteins: ApoE2, ApoE3, and ApoE4. ApoE2 and ApoE3 are considered good because they function as important antioxidants in the brain. In contrast, the ApoE4 variety does not offer any antioxidant protection, which make it less desirable than the other two.

Not everyone carries the same ApoE gene. Each parent donates one ApoE gene to his or her offspring; therefore, each person carries two ApoE genes, and there are six possible combinations. A simple blood test will tell you which genes you carry. Your doctor takes a blood sample and sends it to a special laboratory (Athena Diagnostics), which has the patent for this test. It takes about three weeks to get the results. Ideally, it's best to have the ApoE2 or Apo3 genes in any combination for maximum antioxidant protection. Of course, it doesn't always work out that way, and the lack of antioxidant protection can vastly increase your risk of developing a neurological problem. Carrying even one ApoE4 gene is associated with poorer outcome in surviving traumatic brain injury, earlier age of onset of epilepsy, a higher risk of developing neuropathy if you have diabetes, and significantly faster progression of disability in MS patients. Research also shows that the brain cells of people with one or two ApoE4 genes are deficient in their ability to make energy. The combination of poor antioxidant defenses and an impaired ability to produce enough energy can set the stage for neurological problems.

ApoE4, however, has been most closely correlated with Alzheimer's disease, a condition affecting 4 million Americans that is characterized by the slow and steady destruction of key areas in the brain. Not everyone who gets Alzheimer's has an ApoE4 gene, and not everyone who has the gene will get Alzheimer's, but your risk of getting this disease is significantly higher if you have the gene. For example, if you have only ApoE2 or ApoE3 genes, your risk for developing Alzheimer's by age 80 is 20 percent. But if one of the Apo pairs is E4, the risk increases to 47 percent. If both are ApoE4, the risk is 91 percent.

Is there any point to performing this test at all? I know that many physicians today do not test ApoE status in the mistaken belief that there is nothing that can be done for patients who discover they are carrying the troublesome ApoE4 gene. I recognize that it's not as simple as having a high homocysteine level that can be corrected easily by taking a few vitamins. Nevertheless, I think the "What you don't know can't hurt you" attitude is misguided, especially in light of all we know about free radicals and brain disease. The statistics just cited are based on the experience of patients who were untreated and who never did anything to save their brains. They never changed their diet, took antioxidants or neuronal energizers that improved the production of energy (which increases the ability of brain cells to clean out free radicals), or did anything else to mitigate or correct the underlying problem that put them at risk. I believe that over time, my treatment protocols will have a profound impact on delaying and even preventing the onset of neurological symptoms in patients carrying these genes.

Unfortunately, the ApoE genetic test may not be covered by your insurance company or Medicare, and it could cost you a few hundred dollars out-of-pocket. It is, however, a one-time expense. Your genes don't change, so you don't ever have to redo the test.

Which Tier of the Better Brain Program Is Right for You?

Your test result may affect which tier of the Better Brain program you should follow. If you have *one* ApoE4 gene, it increases your risk level by one tier. If your results on the Brain Audit put you in Tier 1, move up to Tier 2. If your results on the Brain Audit put you in Tier 2, move up to Tier 3.

If you have *two* ApoE4 genes, follow the Tier 3 program for people at highest risk. *This is mandatory!*

Follow-up

Your genes don't change, so there's no point in repeating the ApoE4 test. Your antioxidant status, however, can change for the better. Monitor your progress with the lipid peroxide test. I recommend that you ask your doctor to perform a baseline lipid peroxide test at his or her office and then to do a follow-up lipid peroxide test three months later, after you have been following the Better Brain program. The overwhelming majority of people will see a marked reduction in lipid peroxide levels at that time. A small minority, however, may not. If you don't show any improvement and are following Tier 2, move up to Tier 3 and get retested three months later. If your lipid peroxide levels have not normalized after following Tier 3, switch to Tier 3 Plus (see page 176). If Tier 3 Plus does not get your lipid peroxide levels down to a low range, consider IV glutathione therapy (see Appendix 4).

C-Reactive Protein: Measuring Inflammation

C-reactive protein (CRP) is released by the body in response to acute inflammation. It can be measured by a simple blood test in which a sample is sent to a laboratory to be analyzed. Depending on the laboratory, the results can be available in a matter of days.

What does this test tell you? The higher the level of CRP, the greater the likelihood that your body is being harmed by inflammation. Inflammation is a sign that your brain is under excessive free radical attack and that over time, you will lose more and more brain function. This is precisely how the minor mental lapses that hit in your forties can quickly turn into devastating ones (like a stroke or Alzheimer's disease) in your fifties and beyond. Inflammation doesn't stay in one part of the body. Like a fire raging out of control, it can spread from site to site, causing damage wherever it strikes. In fact, inflammation is considered by some cardiologists to be more of a risk factor for heart disease than elevated cholesterol levels.

Studies have confirmed that the level of CRP in blood, which measures the degree of inflammation in the body, is an excellent predictor of brain disorders such as Alzheimer's disease and stroke. More and more enlightened physicians are offering this test to their patients, and if yours doesn't, you should ask for it or change doctors.

What is normal? A CRP level below 3 milligrams per milliliter is considered optimal and means that your level of inflammation is acceptable.

Which Tier of the Better Brain Program Is Right for You?

Your results on this test may affect which tier of the Better Brain Program you should follow. If your level of CRP is more than *3 milligrams per milliliter,* it increases your risk level by one tier. If, based on your results on the Brain Audit, you are in Tier 1, move up to Tier 2. If your results on the Brain Audit put you in Tier 2, move up to Tier 3.

Follow-up

If you have elevated CRP, it's important to retest every four months after you begin your Better Brain program until your CRP is normal. Some people, however, may never reach a normal CRP. These people may be genetically predisposed to have higher levels of inflammation, which means that they probably also have elevated levels of lipid peroxides. In this situation, it is critical to also closely monitor lipid peroxide levels. Be sure to take the lipid peroxide test (discussed earlier) and follow my protocols to lower your lipid peroxide levels, if need be, and improve your antioxidant status. If your CRP level is stubbornly high but your lipid peroxide levels have been brought down to normal, there is still a substantial reduction in risk.

11 Six Steps to a Better Brain

THE PAGES OF THIS BOOK provide you with all the information you need to maintain a brain that is smart, creative, and flexible. I have shown you how to maintain optimal mental performance and how to keep your brain functioning at its peak so that you can live life to its fullest at any age. I understand that the Better Brain program is a comprehensive approach to brain health that requires time, commitment, and a willingness to make changes in your lifestyle. I know that you may not be ready to commit to following a multifaceted program yet are concerned about protecting your brain against the degenerative forces that will age it (and you) before your time. For those who are willing to do *something,* but not everything outlined in this book, I have highlighted what I consider to be the six most important steps you can take to preserve and protect your brain.

It is my hope that once you get in the habit of following these six basic steps, you will be inclined to follow the entire program.

Step 1: Get the Trans-Fatty Acids Off Your Plate

The first and most important step to make your brain smarter and healthier is to stop eating foods that contain trans-fatty acids. You don't need them, and you will be far better off without them. As you know by now, many brands of margarine and polyunsaturated oils undergo a chemical modification to extend their shelf life and make them easier to use in baking that in the process creates trans-fatty acids. Trans-fatty acids are the primary fats found in packaged baked goods (cookies, cakes, chips) and fried foods. Like any other dietary fat, trans-fatty acids become incorporated into your cell membranes. Unlike healthier fats, trans-fatty acids can make your cell membranes hard and rigid, and will make your brain sluggish. Your ability to remember, learn new things, and maintain a good mood are all dependent on having healthy, flexible cell membranes. Trans-fatty acids will slow your brain response time down while wildly accelerating the aging process.

How do you know if a food contains trans-fatty acids? Read the ingredient labels. Don't buy foods that contain hydrogenated oils or partially hydrogenated oils. Don't eat fried foods—this includes french fries, donuts, and most chips (corn, cheese, and potato). (Baked chips may also contain hydrogenated oils, so read the labels before eating these foods.) Your brain will thank you.

Step 2: Have Your Homocysteine Levels Checked Annually

This really is a no-brainer: ask your doctor to check your homocysteine levels each year as part of your annual physical examination. It's a simple blood test that could alert you to a potential brain problem long before you experience your first symptom. Homocysteine is an amino acid that is produced by every cell in the body and, in excess, can shrink your brain, dull your reflexes (especially those involving hand-eye coordination) and lead to depression. It can also *double* your chances of developing Alzheimer's disease. What is normal? A homocysteine level of over 9 micromoles per liter dramatically increases your risk of neurological problems and should therefore be treated. Anything under 9 is optimal. Elevated homocysteine is easily treated by taking B vitamins. Knowing your homocysteine level is one of the most important things you can do for the health of your brain.

Step 3: Take These Three Supplements

If you can't bear the thought of taking several pills daily, start off by taking three brain-essential supplements. That's three pills, once a day, with food. Really simple. It requires very little commitment, yet the three supplements that I recommend here can help to keep your brain running well right now and protect against brain degeneration down the road. (After you get into the habit of taking these three supplements, I hope that you will be encouraged to follow the entire Better Brain supplement program that is right for you.)

DHA: 300 mg daily
Vitamin E: 200 IU daily (d-alpha, *not* dl-alpha form)
B-complex vitamin: one capsule daily

A basic vitamin B complex supplement should include the following.

B1 (thiamine): 50 mg
B3 (niacin as niacinamide): 50 mg
B6 (pyridoxine): 50 mg
Folic acid: 400 mcg
B12 (cobalamin): 500 mcg

There are several brands of B complex that offer these supplements in one capsule.

Why DHA? About 25 percent of the total human brain fat is composed of DHA; it provides brain cell membranes with the flexibility necessary for efficient communication so that you can think better and faster. This substance is not produced by the body and must be obtained through food or supplements. It's difficult to get enough DHA from food alone, and therefore I recommend that everyone take a DHA supplement. Remember, as I tell my patients, "good fat" makes a great brain.

Why vitamin E? The fat in your brain is especially vulnerable to free radical attack, which is why everyone should take vitamin E daily. Vitamin E is a fat-soluble vitamin, which means that it can get into parts of the cell—notably the cell membrane—that are not accessible to other antioxidants. It's important to remember that when free radicals are allowed to run amok in your brain, they prevent your brain cells from doing their job properly, which is why you become forgetful or have difficulty learning new information or staying focused. The long-term effect of free radicals gone wild can be catastrophic and is linked to Parkinson's disease, Alzheimer's disease, and stroke, among other neurological ailments. Simply taking vitamin E daily can help reduce the lethal effects of free radicals.

Why B complex? If you want to stay in a good mood, you need your B vitamins. If you want to avoid premature brain aging and Alzheimer's disease, you also need your B vitamins. What happens if you don't get enough B vitamins? You are vulnerable to depression, memory problems, and dementia. B vitamins are critical for brain health primarily because they control homocysteine, the amino acid naturally produced by the body. High levels of homocysteine can promote inflammation, damage blood vessels that deliver blood to the brain, and kill brain cells. (That is why I urge everyone to have their homocysteine levels checked each year; see step 2.) In most cases, homocysteine is easily controlled by taking B vitamins, and I urge everyone to take a B-complex supplement.

Step 4: Think Twice Before Taking a Drug

If you are routinely taking a prescription or over-the-counter medicine (three times a week or more) be sure that it is not depleting your brain of important nutrients. Hundreds of drugs sap the brain of B vitamins and antioxidants (like Co-Q10 and glutathione) that are essential for brain health. If you are taking a drug that depletes your brain of important nutrients, it is critical to replenish that nutrient by taking the appropriate supplements. How do you know if you are taking a nutrient-depleting drug? Every time you take a new drug, check the list of nutrient-depleting drugs in chapter 4. If you are, follow the correct supplement program designed to restore those nutrients. This way you'll be sure that you are not inadvertently starving your brain of important nutrients that keep it running smoothly and protect it against premature aging.

Step 5: Get Eight Hours' Sleep Every Night

If you want to function at your mental and physical peak, make sure you get seven to eight hours of sleep on most nights. (On average, we sleep 20 percent less than we did a hundred years ago when nine and a half hours of sleep a night was the norm.) Constant sleep deprivation is very bad for your brain, for both the short term and the long term. Even a mild sleep deficit—getting six hours of sleep for a few nights in a row instead of eight hours—will slow you down and cause notable decline in performance on standard mental function tests. Moreover, when you're tired, you're more likely to be irritable and have less mental and emotional stamina. Tired people misplace things more often, have more trouble recalling information, and in general are not as sharp as those who get enough sleep.

Sleep serves an important purpose for your brain. Sleep is a time when our cells can do vital repair work, "cleaning up" toxins that accumulate in the brain and keeping brain cells in optimal condition. If you don't allow your brain to do this essential housekeeping, toxins will accumulate in your brain over time and accelerate brain aging. Sleep is also a time when your brain processes new information. A good night's sleep can enhance your ability to learn and understand new concepts. In sum, if you don't get enough sleep, your brain will pay the price.

Step 6: Have Some Fun

Make recreational activity a part of your life. It will make you smarter and happier and keep your brain going longer. My only requirement is that you choose an activity that you don't normally do every day as part of your job. Play an instrument. Play bridge. Join a book club. Make pottery. Whatever you choose to do, recreational activity is good for your brain for several reasons. First, it's a great way to relieve stress. When you're concentrating on an activity that you enjoy, it's possible to shut out all the irritants that tie you up in knots and send potentially damaging stress hormones soaring. Second, when you stimulate brain cells with a new challenge, you make new dendrites, the connections between neurons that are essential for the assimilation and processing of information. The more dendrites you have, the better your brain cells can communicate, and the "smarter" you will be. An added bonus: Studies show that people who do some activity beyond work are at lower risk of developing Alzheimer's disease than people who don't. So go out and have some fun. Doctor's orders!

Up until now, I've offered advice on how to prevent brain degeneration and neurological disease. I am confident that the strategies outlined in previous chapters will greatly reduce the number of people who will suffer from neurological problems down the road. And without a doubt, taking the steps to protect your brain now will ensure a better quality of life in the future.

The next part of this book is devoted to people who have already been diagnosed with neurological problems. As you will see, there is new hope for you too.

For the latest information on brain health, visit my website: www.betterbrainbook.com.

Specific Brain
Disorders
and What
to Do About Them

12 Stroke

ABOUT five hundred thousand Americans suffer a stroke each year, making it the most common of all the neurological diseases. Stroke can be debilitating and at times even fatal. About one-third of all stroke victims do not survive the initial attack, making it the third leading cause of death in the United States. There are currently 1.7 million stroke survivors in the United States, but obviously survival does not mean full recovery. Stroke not only is a financial drain—the treatment of stroke patients costs $30 billion annually—but also exacts a steep toll in terms of human suffering. Only 10 percent of stroke survivors can return to jobs and their lives unscathed; 40 percent experience mild disability, and 40 percent are severely disabled, requiring assistance with various tasks of daily living. Sadly, the remaining 10 percent are so disabled that they require nursing home care for the rest of their lives.

It doesn't have to be this way. The Perlmutter Center's protocol for the treatment of stroke victims can vastly improve the prognosis of stroke patients and reduce the amount of long-term disability. Moreover, most strokes need not occur in the first place. Stroke is highly preventable, and simply taking the right vitamins and making adjustments in your lifestyle can remarkably lower your risk.

A stroke is similar to a heart attack, except this time the organ under attack is the brain. Simply put, a stroke occurs when the blood supply to the brain is suddenly cut off and portions of the brain die. As with a heart attack, the most common cause of stroke is a blockage of an artery, interrupting blood flow to a part of the brain. The clot can originate in the carotid artery, the main artery bringing blood into the brain, or it may be caused by a clot breaking free from the heart and traveling upstream. In addition, progressive narrowing of the smaller arteries within the brain can ultimately exceed a critical level, triggering a stroke. A stroke can also occur if a blood vessel weakens and bursts, spilling blood into spaces surrounding the brain cells. This is called a hemorrhagic stroke. In either case, brain cells are killed or severely injured, and this is responsible for the loss of physical or cognitive function.

Stroke increases your risk of another brain-robbing neurological disease—Alzheimer's disease. Many elderly people suffer so-called "silent strokes," in which the blood supply to their brain is compromised for brief periods at a time. Although they may not experience any symptoms, these strokes can be detected on an MRI scan, and they cause a great deal of damage. People who suffer from these events are 2.3 times more likely to develop Alzheimer's disease or another form of dementia than those who do not.

There are three fundamental areas involved in stroke care: *prevention, acute care, and recovery.* Each one is important, although the focus at the Perlmutter Center is on prevention and recovery. Patients in the midst of a stroke—that is, those in need of acute care—are best treated in a hospital setting.

Prevention

The ideal approach to stroke is prevention, that is, reducing or eliminating important risk factors. High blood pressure, elevated blood lipids, smoking, diabetes, obesity, excessive alcohol intake (more than two drinks daily) and a sedentary lifestyle are among the well-known risk factors for both stroke and heart disease. Women who take estrogen—either in the form of oral contraceptives or hormone replacement therapy—are also at greater risk.

For the most part, the same risk factors that increase your odds of coronary artery disease also increase your risk of stroke. Maintaining good cardiovascular health will not only protect you against having a heart attack but will also protect you against having a brain attack.

In recent years, a new culprit has emerged as a leading cause of stroke—elevated homocysteine levels. High homocysteine levels dramatically increase the production of plaque, a mixture of fat and calcified inflammatory tissue that progressively narrows arteries, impeding the flow of blood. Numerous studies have shown a link between high homocysteine and an elevated risk of stroke, especially in the elderly. A recent study conducted at Tufts University showed a direct correlation between high homocysteine and blockage of the carotid artery, one of the main arteries to the brain. In fact, of the 1,041 elderly men and women in the study, those with the highest homocysteine levels had a dramatically increased risk for blockage of the carotid artery. To me the evidence implicating homocysteine is overwhelming, yet the medical community has been painfully slow to act. Once again, I feel compelled to say that it is critical to have your homocysteine levels checked annually. You should demand this test! Elevated homocysteine levels can easily be lowered by taking a B-complex vitamin, and by controlling homocysteine, you greatly reduce the odds of having stroke (and Alzheimer's disease). For more information on the homocysteine test and how to lower elevated homocysteine, see chapter 10.

If you already have had a stroke, it is even more important to maintain a normal homocysteine level, or you increase your risk of having a second stroke. A normal homocysteine level is 9 micromoles per liter or less.

If you are under age 65 and have a primary relative—a parent, grandparent, or sibling—who had a stroke, you are at three times the normal risk of having a stroke yourself. You must be vigilant about maintaining good cardiovascular health and try to avoid taking any medication that may raise your homocysteine level and therefore increase your risk of stroke. (Please review the list of drugs that raise homocysteine in chapter 4.)

Acute Care

Getting the right treatment during the acute phase of stroke—that is, while you're actually having a stroke—can reduce injury to your brain and even save your life. The problem is, many people don't know the signs that they are having a stroke and delay getting medical care. Unlike a heart attack, which often presents with crushing chest pain, a stroke can be more subtle and harder to pinpoint. Many acute stroke patients are unaware that they having a stroke! In fact, some physicians want to change the name "stroke" to "brain attack" so that patients understand the urgency of the situation. If you have any of the symptoms listed below, call 911 for emergency care. Don't try to drive yourself to the hospital, even if you think you can.

- Unexplained dizziness or feelings of unsteadiness
- Loss of vision, especially in one eye
- Loss of speech or difficulty in talking or understanding speech
- Any unexplained weakness or numbness in the face, arms, or legs, or on one side of the body
- A severe headache unlike anything you have experienced before

People may experience one or more of these symptoms very briefly and then resume normal activity. This is called a transient ischemic attack (TIA). It is common to have several TIAs before having a full-blown stroke. Don't ignore a TIA. It is potentially very serious, and you must get immediate medical attention.

If you are in the midst of a stroke, the emergency room physician may administer a drug designed to break up the clot, if appropriate, to help restore normal blood flow to the brain. After a stroke, it's important to detect the cause of the event in order to reduce the risk of it happening again. If your evaluation reveals a severely blocked carotid artery, your doctor may consider surgically opening the blockage. If the clot came from your heart, you may be put on a blood thinner, as well as a program to improve your cardiovascular health, including diet, exercise, and lipid-lowering drugs to normalize cholesterol. It's important to note that many lipid-lowering drugs deplete the body of vital Co-Q10, which is essential for cellular energies. If you are taking one of these drugs, be sure to take supplemental Co-Q10 (see chapter 4).

Recovery

The treatment you get after having a stroke can speed up your recovery and may restore much of your lost brain function. Unfortunately, this critical area has for the most part been neglected by conventional medicine. When a portion of the brain is damaged as a result of trauma, or stroke, a portion of brain tissue is permanently destroyed. The cells have died, and they are now rendered useless. The permanently damaged area is surrounded by another group of cells that, although damaged, still have the potential to be functional once again. These cells are appropriately called "idling" neurons. The Perlmutter Health Center protocol for stroke recovery specifically targets idling neurons, breathing new life into a wounded brain. Although we can't undo all the damage inflicted by a stroke, our therapy is able to restore some metabolic activity to idling neurons. The effects on the patient can be profound in terms of recovering lost function and returning to a more normal life.

The Perlmutter stroke recovery protocol includes hyperbaric oxygen therapy, as well as specific supplements designed to act as cellular energizers that can help reinvigorate flagging metabolic function in brain cells.

Healing with Oxygen

In hyperbaric oxygen therapy (HBOT), patients are exposed to pure oxygen under increased atmospheric pressure. It's painless, safe, and remarkably effective. It is approved by both the FDA and the American

Medical Association for the treatment of a wide variety of medical ailments, including carbon monoxide poisoning, diving injuries such as "the bends," wound healing, and burn therapy. In the United States it is not widely used for stroke rehabilitation, but it is used on virtually every stroke patient in Germany and in many other countries. The use of HBOT for the treatment of stroke is hardly experimental. More than a thousand cases have been reported in medical journals, showing a 40–100 percent rate of improvement in stroke patients given this therapy. In *Stroke,* a highly regarded medical journal, Dr. Richard Neubauer, a pioneer in the use of HBOT for neurological disease, reported outstanding results in a group of 122 stroke patients given HBOT. In one remarkable case, improvement following HBOT was seen in a patient who had had the initial stroke fourteen years earlier!

This therapy works by enhancing the availability of life-giving oxygen to tissues in the body. This boosts oxygen levels in parts of the body with poor or compromised blood supply, as well as in areas in the brain that have been damaged as a result of oxygen deprivation. The infusion of oxygen "wakes up" idling neurons and restores much of their normal metabolic function. The treatment also stimulates the production of new blood vessels, which enhances blood supply to the idling neurons, giving them a major boost in oxygenation and nutrient supply.

Although most patients show immediate benefits even after just one treatment, it can take many weeks of treatments to get the full good result. Typically, the benefits of HBOT for stroke patients include improvement of speech, gait, mental function, and motor power and reduction of spasticity. This treatment doesn't do the job alone—in addition, patients often need physical therapy and/or speech therapy, depending on which part of their brain sustained the damage. In my experience, however, HBOT works in synergy with other forms of therapy, including the specific nutritional supplements described hereafter, to accelerate recovery. It's best to start HBOT as soon as possible after a stroke, but, as noted earlier, cases have been reported of people showing improvements even if they begin therapy years after the stroke. To learn more about the use of hyperbaric oxygen for stroke recovery, visit www.StrokeRecovery.com.

Nutrition

The Better Brain meal plan is good for both your brain and your cardiovascular system. In particular, ridding your body of bad fats (trans-fatty acids and saturated fats) is key to maintaining brain and heart health. These bad fats not only promote neurological disease but accelerate atherosclerosis or hardening of the arteries, making you more vulnerable to a heart attack or a brain attack.

A low intake of B vitamins, notably folic acid, B6, and B12, is associated with an elevated risk of stroke, undoubtedly because of elevated homocysteine levels. Whole, unprocessed grains, leafy green vegetables, and lean protein are excellent sources of these vitamins.

Stroke risk is markedly reduced in people who consume fish, even just once a month. Most researchers credit this risk reduction to the important essential fatty acids found in fish, notably DHA. That is why my nutritional protocols include these important essential fatty acid supplements.

A diet rich in antioxidants can also protect against stroke and heart disease. Free radicals injure the arterial wall, promoting the formation of plaque that blocks the flow of blood and oxygen. Antioxidants, such as vitamin E and vitamin C, help to keep free radicals under control and your arteries clear.

Obesity is a major risk factor for heart disease and stroke. I believe that the obesity epidemic is due to poor diet in general and the overconsumption of commercial, processed foods in particular. Eliminating processed foods from your diet will help you shed unwanted pounds fairly quickly and will dramatically decrease your risk of all chronic disease.

Anyone who has had a stroke or a TIA should follow the Tier 3 supplements regimen (see chapter 6 and Appendix 1).

If your stroke was due to a brain hemorrhage, do not use vinpocetine. (For more information on vinpocetine, see chapter 6.)

Supplements Regimen for Stroke Patients

SUPPLEMENT	A.M.	P.M.
DHA	300 mg	300 mg
Co-Q10	100 mg	100 mg
Vitamin E	400 IU	
Vitamin C	500 mg	500 mg
Alpha lipoic acid	200 mg	
N-acetyl-cysteine (NAC)	400 mg	400 mg
Phosphatidylserine	100 mg	50 mg
Acetyl-L-carnitine	400 mg	400 mg
Ginkgo biloba	60 mg	
Vitamin D	400 IU	
Vinpocetine*	5 mg	5 mg
Vitamin B-complex supplement:**		
B1 (thiamine)	50 mg	
B3 (niacin as niacinamide)	50 mg	
B6 (pyridoxine)	50 mg	
Folic acid	400 mcg	400 mcg***
B12 (cobalamin)	500 mcg	500 mcg***

*Vinpocetine is required only for people with high homocysteine or a history of vascular dementia or coronary artery disease. It should be avoided by people taking Coumadin (warfarin), a blood thinner. Vinpocetine should not be used by people who have had a hemorrhagic stroke.

**Look for a B-complex supplement containing several B vitamins in one capsule or pill.

***In addition to your B complex in the A.M., take an additional 500 mcg of B12 and 400 mcg of folic acid in the P.M.

13 Vascular Dementia

VASCULAR DEMENTIA is the second most common cause of dementia—only Alzheimer's disease claims more victims. Although Alzheimer's disease and vascular dementia are different problems, there is often overlap of these two diseases. In fact, many people with Alzheimer's disease also have some degree of vascular dementia, which has led many researchers to speculate that vascular dementia may be an underlying cause of Alzheimer's disease. Moreover, stroke victims are at increased risk of vascular dementia and Alzheimer's disease, which demonstrates a link between these three brain disorders. Much of what I say about preventing Alzheimer's disease and stroke is applicable to vascular dementia.

Like stroke, vascular dementia is a problem that originates in the cardiovascular system. Vascular dementia is due to damage to the small blood vessels supplying blood to the brain. As these vessels become less viable and blood flow is reduced, patients experience a decline in mental function. Unlike a stroke, which can cause immediate disability and even death, the effects of vascular dementia are more often subtle yet over time can be very damaging.

Symptoms of vascular dementia include depression, confusion, migraine-like headaches, emotional problems such as uncontrollable laughter or crying, inability to handle money, and loss of bladder or bowel control. Vascular dementia typically strikes between the ages of 60 and 75 and is more common in men than women.

Although vascular dementia sounds a lot like Alzheimer's disease, there are important differences. First, while the course of Alzheimer's disease is slowly progressive, vascular dementia is often characterized by a stepwise progression of symptoms that can worsen significantly even over the course of just a few days. Second, vascular dementia is much more frequently associated with other vascular problems like coronary artery disease and high blood pressure, which is why it is so important to maintain normal blood pressure. Both Alzheimer's disease and vascular dementia are strongly associated with diabetes.

Elevated homocysteine is another critically important risk factor for vascular dementia. Tragically, most clinicians do not appreciate this important relationship. Simply checking homocysteine and using appropriate B vitamins to normalize it when elevated could spare thousands of elderly people from dementing illness.

People who have been diagnosed with vascular dementia should follow the Tier 3 supplements regimen (see chapter 6 and Appendix 1). In addition, I recommend the supplement *vinpocetine*. An extract of the periwinkle plant, vinpocetine can increase blood flow to the brain, reduce blood stickiness or the risk of blood clots, increase the brain's ability to make energy, and act as a potent antioxidant. Studies have clearly demonstrated that it is an effective treatment for vascular dementia. In one Italian study, researchers noted that the majority of patients taking vinpocetine made "good to excellent improvement since entering the study" and "scored consistently better than placebo patients in all evaluations in the effectiveness of treatment." I have found the same good results in my patients.

Supplements Regimen for Vascular Dementia Patients

SUPPLEMENT	A.M.	P.M.
DHA	300 mg	300 mg
Co-Q10	100 mg	100 mg
Vitamin E	400 IU	
Vitamin C	500 mg	500 mg
Alpha lipoic acid	200 mg	
N-acetyl-cysteine (NAC)	400 mg	400 mg
Phosphatidylserine	100 mg	100 mg
Acetyl-L-carnitine	400 mg	400 mg
Ginkgo biloba	60 mg	
Vitamin D	400 IU	
Vinpocetine*	5 mg	5 mg
Vitamin B-complex supplement**		
B1 (thiamine)	50 mg	
B3 (niacin as niacinamide)	50 mg	
B6 (pyridoxine)	50 mg	
Folic acid	400 mcg	400 mcg***
B12 (cobalamin)	500 mcg	500 mcg***

*Vinpocetine is required only for people with high homocysteine or a history of vascular dementia or coronary artery disease. It should be avoided by people taking Coumadin (warfarin), a blood thinner. Vinpocetine should not be used by people who have had a hemorrhagic stroke.

**Look for a B-complex supplement containing several B vitamins in one capsule or pill.

***In addition to your B complex in the A.M., take an additional 500 mcg of B12 and 400 mcg of folic acid in the P.M.

Alzheimer's Disease

OF ALL THE ailments associated with aging, the thought of Alzheimer's disease is the one that strikes the most terror in people's hearts. It is a progressive, debilitating disease that can rob you of your ability to communicate, think clearly, and function. You lose your awareness of yourself and your environment, and what's worse, you lose all control over your life. Alzheimer's puts an enormous financial and emotional strain on families, who must care for patients 24 hours a day. Many people with Alzheimer's eventually end up in a nursing home because they can't perform even the simplest tasks, like dressing, feeding themselves, or taking care of personal hygiene. I'm not going to tell you that I have a cure for this cruel disease. As yet there is no cure, and conventional medicine offers limited treatments to deal with symptoms. I can report, however, some success in dramatically slowing down the progression of this disease. I have treated patients who, after a year or two of receiving their diagnosis, have not experienced significant deterioration and are still able to function. My Alzheimer's patients follow a very aggressive antioxidant, anti-inflammatory supplement protocol that gets to the underlying cause of Alzheimer's disease—free radical–mediated deterioration in the brain, and inflammation. Moreover, I frequently see actual modest improvements in memory and cognitive function in patients who follow my program, especially if we begin intervention early.

More important, public awareness needs to radically increase that in many cases this devastating disease can be *prevented.* Alzheimer's disease can be substantially reduced by lifestyle changes that include exercise, maintaining normal weight, leisure activities, stress reduction, antioxidant consumption, and other factors (see chapter 10). Unfortunately, our society focuses on revenue-generating pharmaceutical intervention for established medical problems. With all the published science appearing in the medical literature describing methods of Alzheimer's prevention, it is sad that until now this information has not been widely known by the public. Sadly, the generation of current Alzheimer's patients did not have access to this information; nevertheless, they can still be helped.

Alzheimer's disease is the primary cause of dementia among the elderly. There are 4.5 million Americans with Alzheimer's disease, and by 2030, as the baby boom generation ages, this number is expected to grow to 9 million. About 10 percent of all people over 65 will get Alzheimer's, but if you live to be 85, you have a 50 percent chance of developing this disease. It may seem crass to equate human suffering with money, but the fact is that Alzheimer's disease is a major drain on the economy. Right now it costs about $60 billion a year to care for the current group of Alzheimer's patients. If we could delay the onset of Alzheimer's by just five years—which I believe could easily be achieved—we could cut that cost in half *and* give Alzheimer's victims the gift of five disease-free years.

What Is Alzheimer's Disease?

Alzheimer's disease is characterized by the accumulation in the brain of a protein called beta amyloid. Beta amyloid is a substance that is normally found in the body, but at around age 50, it can start to accumulate in plaques in the brain. Over time, beta amyloid plaques increase free radical production and inflammation, which kill neurons, resulting in a loss of memory, loss of control of body function, and ultimately death. The hippocampus, the memory center of the brain, is especially vulnerable to plaque-inflicted damage, which is why one of the first signs of Alzheimer's is memory loss.

Alzheimer's patients are also profoundly deficient in an important neurotransmitter, acetylcholine, which is involved in memory function and may also help to preserve nerve cell membranes, including the synapses (the communication points between the brain cells) intact. This enables brain cells to talk to each other and share information. As we age, we naturally produce less acetylcholine, but the drop in production of this important neurotransmitter is much steeper among Alzheimer's patients.

The drop in production of acetylcholine is probably a *result* of damage to neurons and not the *cause*. As plaque spreads through the brain, it destroys specific areas involved in memory and behavior. This destructive process can progress for years, sometimes decades, before Alzheimer's victims experience their first symptoms.

Diagnosing Alzheimer's Disease

There is no one specific test to diagnose Alzheimer's disease. The only definitive diagnosis of Alzheimer's can be made by an autopsy, which will reveal the telltale amyloid plaque deposits in the brain. Brain scans may show shrinkage of the hippocampus and of other brain structures, but this does not provide a definitive diagnosis. Instead, physicians rely on cognitive function tests, an interview with the patient, a careful medical history, and family members' assessment of a patient's mental status.

The most common symptoms of Alzheimer's include:

* Short-term memory loss that worsens over time; patients may forget names of people they know and appointments
* Difficulty speaking or finding the right words to express oneself
* Forgetting how to use common, everyday objects (like a pencil or a fork)

- Forgetting to do routine tasks such as turning off the stove, locking the door, or closing the windows
- Mood changes such as irritability, depression, and difficulty paying attention
- An inability to reason or to solve everyday problems that were not previously challenging
- Signs of disorientation, being unable to drive, or feeling "lost" in familiar surroundings
- As the disease progresses, episodes of delusion

Some of these symptoms can occur in people who do not have Alzheimer's. We've all misplaced objects, temporarily forgotten a name or a phone number, or experienced moodiness. The difference is that when you have Alzheimer's these symptoms worsen over time, and you may experience them more frequently with greater intensity. As I tell my patients, if you think that something is wrong with your brain function, it's a *good* sign. It means that you still have the cognitive ability to discern a problem and that your disease is still in its earliest and most treatable stages.

Symptoms may progress faster in some people than in others, but the average patient lives 7–10 years after the initial diagnosis. The usual cause of death is a lung infection such as bronchitis or pneumonia.

If you have any signs of Alzheimer's, it's important to check with your physician to rule out other medical problems, such as low thyroid, heart disease, infection, hearing problems, vitamin B12 deficiency, or even severe depression, which can all cause symptoms that are similar to Alzheimer's disease. In these cases, symptoms disappear once the problem is properly treated. Moreover, many drugs can cause memory loss, confusion, and other Alzheimer's-type symptoms that vanish once the drug is discontinued. I have "cured" many cases of so-called dementia simply by telling patients to discontinue taking certain medications!

If you go to a conventional doctor, he or she will probably prescribe drugs (which I describe later) that are generally ineffective. You will then probably be told to get your affairs in order. It's a good idea for you to plan for the time when you may be unable to care for yourself, but I believe my protocol can buy you time, help you regain your mental function, and extend your life.

Getting to the Underlying Cause

Everybody has beta amyloid in their bodies, but what causes the excessive growth of beta amyloid plaque in the brains of Alzheimer's victims? There is compelling evidence that free radicals play a major role. Free radicals not only trigger the growth of plaque, but plaque itself enhances the proliferation of free radicals. This creates a tremendous burden on the brain, which is low in protective antioxidants under the best of circumstances. (Alzheimer's victims often have the added disadvantage of producing lower-than-normal levels of antioxidants.) In addition, elevated blood sugar directly increases the amount of beta amyloid plaque in the brain. Not surprisingly, diabetics are at much higher risk for Alzheimer's disease. Beta amyloid plaque also promotes inflammation, which only intensifies free radical production, enhancing the damage to delicate brain tissue.

At this point, there is no "official" cause of Alzheimer's disease that is accepted by the medical community, but I believe that we know enough about the underlying process of this disease to make some scientifically based conclusions.

Too few antioxidants, too many free radicals. Alzheimer's patients typically do not have strong antioxidant defenses to protect their brains against free radical attack. Some, but not all, Alzheimer's patients carry a variation of a gene known as ApoE, that gives them less antioxidant protection than normal. ApoE is a protein involved in cholesterol transport, and you can inherit one of three versions of this gene from your parents. ApoE2 and ApoE3 are considered good because they provide important antioxidant protection for the brain. ApoE4 does not offer any antioxidant protection. If you carry an ApoE4 gene, you are at greater risk of developing Alzheimer's disease, and at an earlier age, than if you carry the other types of ApoE genes. This is compelling evidence that antioxidants can protect against Alzheimer's, suggesting that free radicals are clearly playing a pivotal role in this disease.

There's yet more evidence pointing to the power of antioxidants as protectors against Alzheimer's and, indirectly, incriminating free radicals. In a recent study reported in the *Archives of Neurology* (June 2002) researchers found a direct correlation between high levels of lipid peroxide (byproducts of free radical production) and mild cognitive impairment (MCI) in elderly patients. About half of all people with this problem will go on to develop Alzheimer's. What was the primary difference between the patients who went on to develop Alzheimer's and those who did not? Those who developed Alzheimer's had higher levels of free radical activity than those who did not, which indicates poor antioxidant protection. On the basis of their observations, these researchers concluded that oxidative damage to the brain is "an early event in the development of Alzheimer's type pathology." Studies like this leave no doubt as to the critical importance of antioxidants in protecting the brain, and knowing your antioxidant status by checking your lipid peroxide level (see chapter 10).

High homocysteine. Numerous studies have linked elevated homocysteine to dementia, Alzheimer's disease, and poor performance on cognitive function tests. As noted in a recent study in the *New England Journal of Medicine* (February 14, 2002), "an increased plasma homocysteine level is a strong, independent risk factor of the development of dementia and Alzheimer's disease." This is an extremely powerful and important statement coming from what is perhaps the world's most respected medical journal. Do you know your homocysteine level? See chapter 10 to find out more about this important test.

Other studies have confirmed that high levels of homocysteine can cause direct damage to the hippocampus, the memory center of the brain that is most affected by Alzheimer's disease. What do these findings tell us about the role homocysteine plays in Alzheimer's disease? Elevated homocysteine levels are associated with high levels of inflammation, which is believed to play a role in the formation of amyloid plaque in the brain. Inflammation also generates more free radicals. (Every time I mention elevated homocysteine, I feel compelled to say that it is easily controlled by taking a few B vitamins every day. It is tragic that so many physicians do not test or educate their patients about this problem, leaving them vulnerable to Alzheimer's disease and other life-threatening conditions.)

High inflammatory markers. The CRP test that I recommend in chapter 5 is a marker for inflammation. Based on a recent study, men with the highest levels of CRP at midlife—long before the onset of any clinical symptoms—had triple the risk of developing dementia or Alzheimer's disease later in life compared to men with the lowest levels of CRP. This begs the question: *What causes inflammation in the brain?* Homocysteine is one culprit, but exposure to toxins is probably another. For example, there are high levels of aluminum in the brains of Alzheimer's patients, and like other metals, aluminum can promote inflammation and promote the formation of free radicals.

Is Your Painkiller Hurting Your Brain?

People who routinely take nonsteroidal anti-inflammatory drugs (such as ibuprofen) to treat unrelated conditions such as arthritis are at lower risk of Alzheimer's disease than people who do not take anti-inflammatory drugs. In contrast, people who routinely use acetaminophen, a common over-the-counter pain reliever, are at higher risk of Alzheimer's disease. Why? Acetaminophen lowers the level of glutathione, a powerful antioxidant that protects the brain against free radicals.

It's interesting to note that this same process—the free radical–inflammation cycle—is similar to the one that causes arthritis. In fact, the same inflammatory chemicals found in the brain are also found in arthritic joints. They cause the joints to stiffen up and are responsible for the aches and pains of arthritis. The difference is, arthritis victims know that they have a problem because they are able to feel pain. What makes Alzheimer's so insidious is that the brain doesn't have pain receptors. *The brain can be slowly destroyed without our even knowing it until we begin to lose brain function.*

Conventional Drug Therapy

There are very few conventional treatments for Alzheimer's disease and only one that I would consider to be effective—memantine (brand name: Namenda)—a new drug that was developed in Germany and recently brought to market in the United States. It's the only drug to date that has been shown to produce noticeable improvement in advanced Alzheimer's patients, reducing their need for caregivers. Patients who took memantine were better able to perform the activities of daily living (like dressing, washing, and eating by themselves) and were also better able to recognize family members and carry on simple conversations. Unlike other drugs used to treat Alzheimer's, memantine has no known side effects. At the Perlmutter Health Center, we had the opportunity to try this drug on several dozen of our Alzheimer's patients long before it became available in the United States. It was definitely helpful but hardly a cure. It does not help people in the earliest stages of the disease, but considering the total lack of treatment for people with advanced Alzheimer's, it is a very encouraging sign.

Memantine works by blocking the action of the amino acid glutamate, which has been linked to free radical production in the brain. The good news is, the development of this drug shows that pharmaceutical companies are finally paying attention to the critical role that free radicals play in brain degeneration diseases. To me, this underscores the importance of following a *preventive* program to prevent free radicals from wreaking havoc on the brain before Alzheimer's sets in.

There are four other drugs commonly used to treat Alzheimer's disease: Reminyl (galantamine); Exelon (rivastigmine); Cognex (tacrine), and Aricept (donepezil). These drugs all work by boosting levels of acetylcholine, the neurotransmitter that is markedly low among Alzheimer's patients. All of these drugs have significant side effects and, in my opinion, are often used inappropriately. While the literature supporting these drugs indicates that they may produce a mild improvement in symptoms, their benefits are typically unnoticed by patients and family members. Moreover, serious side effects are seen in many patients using these medications, including liver failure, urinary obstruction, seizures, and even death. Paradoxically, these drugs may lead to "sudden worsening." I am not convinced that the benefits of these drugs outweigh the potentially damaging effects.

Nutrition Advice

Whether you have already been diagnosed with Alzheimer's disease or you want to prevent it, the Better Brain meal plan is designed with you—and the health of your brain—in mind. You should follow the basic plan but be vigilant about adhering to the following guidelines.

Saturated fats and trans fats are your brain's enemy. People who consume high amounts of saturated fats (found in meat and whole-fat dairy products) and trans-fatty acids (found in fried foods, many brands of margarine, and processed baked goods) are at greater risk of Alzheimer's disease. Limit your intake of saturated fats to no more than 10–15 percent of your total daily calories to reduce your risk. Use only the leanest cuts of meat and watch your portion sizes. White meat chicken, turkey, and fish are lower in saturated fat than beef and lamb.

Good fats can reduce your risk of Alzheimer's disease. Use olive oil, walnut oil, canola oil, and pumpkinseed oil in your cooking and on your salads.

Eat vitamin E–rich foods. People who consume the most vitamin E throughout their lives, either through diet or supplements, are at the lowest risk of dementia. The good fats just mentioned contain high amounts of vitamin E. Nuts, nut butters, whole grains, and leafy green vegetables are also excellent sources of this brain-saving vitamin.

Supplements

At the Perlmutter Center, our Alzheimer's disease protocol focuses on reducing free radical attack and cooling down the inflammation that is destroying delicate brain tissue. Our goal is to keep our patient functioning well, and the symptoms at bay, for as long as possible. We are vigilant about maintaining normal homocysteine levels and checking lipid peroxide levels (with the lipid peroxide test described in chapter 10) to make sure that we are providing ample antioxidant protection to keep free radicals under control. I have found a consistent and direct correlation between elevated lipid peroxide levels (meaning high free radical activity) and progression of symptoms, not only in Alzheimer's but in all degenerative diseases of the brain. Once we bolster their antioxidant defenses with appropriate supplements, very often patients are rewarded with improvements in mental function, indicating a slowing down in the progression of their disease.

You should follow the Tier 3 supplements regimen (see chapter 6 and Appendix 1). I also recommend two additional supplements.

Borage oil. This is a great source of GLA, an essential fatty acid that can help reduce brain inflammation. Take a dose that supplies about 300 milligrams of GLA daily. You can find borage oil in most health food stores.

Melatonin. A hormone naturally produced by the pineal gland in the brain, melatonin is best known as the hormone that regulates the sleep-wake cycle. Melatonin is also an excellent antioxidant and protects the hippocampus, the brain structure that is vital for memory function. Take 3–9 milligrams of melatonin at bedtime. Begin with 3 milligrams, and if you still can't fall asleep, increase your dose by 1 milligram every half hour or so until you get to a maximum of 9 milligrams.

Supplements Regimen for Alzheimer's Patients		
SUPPLEMENT	A.M.	P.M.
GLA (borage oil)	300 mg	
Melatonin		3–9 mg at bedtime
DHA	300 mg	300 mg
Co-Q10	100 mg	100 mg
Vitamin E	400 IU	
Vitamin C	500 mg	500 mg
Alpha lipoic acid	200 mg	
N-acetyl-cysteine (NAC)	400 mg	400 mg

Phosphatidylserine	100 mg	100 mg
Acetyl-L-carnitine	400 mg	400 mg
Ginkgo biloba	60 mg	
Vitamin D	400 IU	
Vinpocetine*	5 mg	5 mg
Vitamin B-complex supplement:*		
B1 (thiamine)	50 mg	
B3 (niacin as niacinamide)	50 mg	
B6 (pyridoxine)	50 mg	
Folic acid	400 mcg	400 mcg**
B12 (cobalamin)	500 mcg	500 mcg***

*Look for a B-complex supplement containing several B vitamins in one capsule or pill.

**In addition to your B complex in the A.M., take an additional 500 mcg of B12 and 400 mcg folic acid in the P.M.

***Vinpocetine is required only for people with high homocysteine or a history of vascular dementia or coronary artery disease. It should be avoided by people taking Coumadin (warfarin), a blood thinner.

Parkinson's Disease

THE PERLMUTTER HEALTH CENTER'S treatment of Parkinson's disease is a true model of complementary medicine. Our approach is to combine the best treatments from both conventional medicine and alternative medicine to create a better model of care for patients. Without question, the pharmaceutical industry has provided powerfully effective drugs for the treatment of this disease, and we incorporate medications into our protocol when they are clearly needed. We prescribe some of the standard drugs for Parkinson's disease, but we also incorporate an intravenous glutathione therapy, targeted supplements, and nutritional counseling. The results are nothing short of amazing. Although we can't cure Parkinson's disease, the overwhelming majority of patients vastly improve on our Parkinson's regimen. Patients who could barely drag themselves along with walkers are now walking—some are even running—with relative ease. Patients who were debilitated by embarrassing tremors are now in greater control of their movements. Patients who were slurring their words are now better able communicate with their friends and loved ones. Patients who had difficulty performing the most mundane activities of everyday life are now self-sufficient. To me and to my patients and their families, these improvements are miraculous.

Unfortunately, this is not the reality for most Parkinson's patients. The current mode of treatment is sorely inadequate and obsolete. The sad truth is that the standard drug treatment for Parkinson's merely masks the symptoms and does nothing to stop the progression of the disease. In some ways it may actually make it worse. In contrast, the Perlmutter Center's treatment not only significantly relieves symptoms but actually slows down the underlying process that is causing this disease.

About 1.5 million Americans have Parkinson's disease, and some 50,000 new cases are diagnosed each year. Men are more vulnerable to Parkinson's disease than women, although no one knows why. In recent years, several high-profile people with Parkinson's disease have raised the public's consciousness about this disease, including actor Michael J. Fox, former attorney general Janet Reno, prizefighter Muhammad Ali, and writer Michael Kinsley. What is striking is that all of these people developed Parkinson's well before 60, the average age of onset. Unfortunately, they are not alone. Though it was once rare among the young, scientists estimate that around 5–10 percent of all cases of Parkinson's now occur in younger people, some in their twenties. Not only is the incidence of Parkinson's, like that of other neurological diseases, on the rise, but the disease is hitting people at younger and younger ages. In previous chapters, I've discussed some of the reasons why the incidence of neurological disease is increasing, even among the young who should be resistant. Our increasingly toxic environment is lethal to the nervous system, so it is no surprise that we are seeing a surge in neurological disorders.

What Is Parkinson's Disease?

More than five thousand years ago, a mysterious ailment with symptoms similar to those of Parkinson's appeared in the medical texts of the ayurvedic healers of India. Through the ages, other medical texts mentioned symptoms or groups of symptoms that could have been Parkinson's, but since Parkinson's can manifest itself in so many different ways, it was difficult for ancient healers and later physicians to connect the dots. The disease wasn't identified as a distinct ailment until 1817, when Dr. James Parkinson first described this condition in his article "An Essay on the Shaking Palsy."

We now know that Parkinson's is a progressive neurological disorder that results from the degeneration of the dopamine-producing area of the brain, the substantia nigra (black substance). Dopamine is the neurotransmitter that helps control balance and movement. Parkinson's patients have significantly lower levels of dopamine than normal.

Getting to a Diagnosis

There is no one medical test to diagnose Parkinson's disease. It is a diagnosis that is based on symptoms, family history, and personal medical history. Bran scans can help eliminate other neurological problems but do not provide specific diagnosis of Parkinson's. The brains of Parkinson's patients do have a hallmark structure, called Lewy bodies, that distinguishes this brain disease from others; however, Lewy bodies are only revealed after death if an autopsy is performed. If a physician suspects that a patient has Parkinson's, the usual protocol is to give him or her an anti-Parkinson's drug, such as L-dopa, to see if there is any change in symptoms. After taking a dopamine-boosting drug, Parkinson's patients typically experience a relief in symptoms, at least for a few hours, which helps confirm the diagnosis.

The Symptoms

As I tell my patients, there is no typical case of Parkinson's disease. Symptoms vary widely from patient to patient, and the disease can progress slowly or rapidly. Just because a friend or a family member may have fared poorly doesn't mean that you are destined to follow in his footsteps, and the odds of you doing well are greatly enhanced if you get proper treatment.

Symptoms common to most Parkinson's patients include a distinctive tremor, muscle rigidity, slowness of movement, and disturbances of posture. Parkinson's is perhaps best known for its characteristic tremor, which often begins in one hand may spread up the arm and eventually to the legs, facial muscles, and tongue. It could stay on one side or spread to both sides of the body. Unlike non-Parkinson's tremors, the Parkinson's tremor may get worse at rest and often improves when the hand or leg is in use. Fortunately, the tremors are usually not disabling and generally disappear during sleep.

Slowness of movement, medically known as bradykinesia, is another hallmark of Parkinson's disease. Parkinson's patients develop a characteristic slow gait, which is more a shuffle than a walk. Many Parkinson's patients can't walk without the aid of a walker, and some may require a wheelchair. As the disease progresses, posture is often affected, and the Parkinson's patient walks in a stooped, forward position. Muscle stiffness occurs in the limbs and neck. At times muscles may freeze up, making it hard to do simple things such as get up from a chair or start to walk from a standing position. Patients often describe the sensation as feeling as if they are wearing cement boots or as if their feet are magnets stuck to a metal surface. The handwriting of people with Parkinson's is often strained, becoming smaller and smaller and increasingly difficult to read. There is also a loss of automatic movements, that is, you no longer can do the things that you used to do unconsciously, such as blinking your eyes, smiling, or swinging your arms when you walk. What's particularly frustrating to both patients and loved ones is that facial expressions are reduced, so patients can't fully express their feelings. Despite the fact that inside they may still feel and think with the same passion as before they were afflicted with this disease, Parkinson's patients often have a vacant, empty stare as a result of the inability to spontaneously move facial muscles. Moreover, Parkinson's patients often have difficulty speaking, and the voice may become soft and expressionless.

Complications from Parkinson's disease can include difficulty chewing, urinary incontinence, constipation (which is often a result of medicine given for Parkinson's), insomnia, and loss of libido or sexual dysfunction. About 30 percent of Parkinson's patients lose some of their cognitive function as the disease progresses. Depression is also very common among Parkinson's patients, which may be part of the natural disease process, a side effect of medication, or a natural reaction to the stress of living with a chronic, often debilitating ailment.

The Underlying Cause

Parkinson's disease is characterized by a decreased production of dopamine as a result of the degeneration of the neurons in the dopamine-producing section of the brain. This begs the question: *What causes the brain to self-destruct in the first place?* In recent years, scientists have discovered that the brain cells of Parkinson's patients are not able to produce enough energy for normal activity. The slowdown in energy production occurs when free radicals attack healthy cells and destroy the mitochondria, the energy-producing centers of cell membranes. Without adequate energy, these cells cannot produce enough dopamine. As dopamine production falls off, the symptoms become more severe.

When discussing risk factors for neurological diseases (see chapter 3) I noted that people who have been exposed to particular toxins, such as insecticides used in farming or gardening or other chemicals on the job, are prime candidates for Parkinson's disease. This makes perfect sense, because toxins increase the production of free radicals in the body, particularly in the brain. It's true, however, that not everyone who is exposed to these toxins will eventually develop Parkinson's disease. Why are some people more vulnerable to the damaging effects of toxin-generated free radicals than others? Interestingly, the answer may have more to do with the liver than the brain. The liver is the body's main detoxifying organ. Its primary job is to render chemicals harmless before they can enter the bloodstream and circulate throughout the body's tissues and cells. Many Parkinson's patients have weaker-than-normal liver function and therefore have a flawed detoxification system. When I suspect that a patient has Parkinson's, I often recommend that he or she undergo a test called a hepatic detoxification profile, offered by Great Smokies Diagnostic Laboratory in Asheville, North Carolina. The test involves the oral administration of several over-the-counter challenge substances, including caffeine, aspirin, and acetaminophen. Saliva, blood, and urine are collected to determine how well these substances are metabolized. This test provides an extremely comprehensive picture of the various liver detoxification pathways, allowing the physician to design a specific program to enhance liver function. If a patient's test reveals a general weakness in his detoxification system, I recommend a combination of specific foods and supplements designed to bring liver function up to speed. Typically, the program provides higher amounts of sulfur-containing foods, such as cruciferous vegetables (broccoli, cauliflower, Brussels sprouts, kale), which provide the raw materials for glutathione, and supplements such as milk thistle (one 200 mg capsule twice daily) that help boost liver function. You can find milk thistle at most health food stores and many pharmacies. It's also important to avoid alcohol, which can further burden the liver. (Intravenous glutathione therapy, described in Appendix 4, can also help improve liver function.)

Once you understand the actual cause of the decrease in dopamine, the thought of merely boosting dopamine levels and ignoring the underlying process damaging the brain, which is the standard mode of treatment, is tantamount to tending to the smoke and not the fire. In order to treat the Parkinson's patient effectively, you must do both.

Treatment: The Conventional Approach

The standard drug therapy for Parkinson's is levodopa, or L-dopa, marketed as Sinemet. If used correctly, it is very effective at reducing the symptoms of Parkinson's, at least initially. L-dopa, however, has its shortcomings. After a while, patients may develop a tolerance to the drug and need to take higher and higher doses to get the same beneficial effect. Moreover, patients taking L-dopa for Parkinson's have higher-than-average levels of homocysteine, which increases the risk of dementia, stroke, and heart disease. In fact, some researchers have questioned whether the drug-induced rise in homocysteine is related to the increased incidence of dementia among Parkinson's patients. I do use L-dopa in my practice, but I am able to keep my patients on relatively low levels because I use L-dopa in conjunction with other therapies. In addition, I am very careful to closely monitor the homocysteine levels of my patients and to adjust their intake of B vitamins accordingly. If you are taking L-dopa, you must have your homocysteine level checked. If your level is high (greater than 9) you should begin a program to lower homocysteine (see chapter 10 for more details).

Drugs called dopamine agonists are used to enhance the effectiveness of L-dopa. These include Mirapex (generic name: pramipexole) and Requip (generic name: ropinirole.) Although these dugs may be somewhat helpful in controlling symptoms, they are associated with drowsiness and, in some cases, hallucinations.

There are a variety of other medications used to treat the symptoms of Parkinson's disease. Keep in mind, however, that all of these medications have potential downsides, so strive for the lowest dosage possible to achieve the goal of symptoms management.

Surgical Options

As many Parkinson's patients are aware, there are now several surgical procedures being utilized for treating the symptoms of the disease. The two most common types of surgery involve either the electrical stimulation or the brain or destroying a portion of the brain affected by Parkinson's.

In the electrical stimulation procedure, an electrode is implanted in a specific part of the brain and is connected to a pacemaker device located in the chest or abdomen. Electrical stimulation is sent to the specific area of the brain and can be adjusted for maximal benefit. We have used this procedure in a small number of our patients and it has proven clearly helpful in reducing tremor and, to a lesser degree, rigidity in most of these people. Keep in mind, however, that this is a significantly invasive procedure accompanied by risks of permanent injury or even death.

The second surgical procedure involves destroying a specific area of the brain thought to be overactive in Parkinson's disease. Like the electrical stimulation procedure, this technique offers risks as well as benefits. Unlike placing an electrode in the brain, which can be removed, this procedure requires the destruction of a particular structure in the brain and is irreversible.

While there are patients who can benefit from these procedures, in my opinion, their numbers are few. Clearly, these surgical techniques have their place, but only after less risky medical and nutritional therapies have failed.

The Glutathione Miracle

Glutathione is an antioxidant produced by the body that is vital for the health of your brain and is also critical for proper liver function. It is sorely deficient in patients with Parkinson's disease, a fact that is overlooked by the conventional treatment for this disease, which has focused almost exclusively on boosting dopamine availability. In 1996, while I was searching for a more effective treatment for Parkinson's disease, I came across a study conducted in Italy in which Parkinson's patients were given intravenous glutathione twice daily for 30 days. The participants were evaluated at one-month intervals up to six months. The authors noted that "all patients improved significantly after glutathione therapy, with a 42 percent decline in disability. Once glutathione was stopped, the therapeutic effect lasted 2–4 months." The researchers reported virtually no side effects with the treatment.

Secure in the knowledge that glutathione was safe, and appeared to be even more effective than current established therapies, I wanted to try it on my own patients. In 1998 the Perlmutter Health Center began administering intravenous glutathione to patients. It is a true miracle "drug." Following even a single dose of glutathione, patients typically experience a rapid improvement in symptoms, often in as little as 15 minutes. Our Parkinson's patients are now realizing profound improvements with respect to reduction in rigidity, increased mobility, improved speech, better mood, and decreased tremor. I have seen wheelchairbound patients walk again after a few treatments, and I've seen patients whose faces were frozen in blank stares smile and laugh again. I've literally seen people brought back to life.

Glutathione, being perhaps the most powerful brain antioxidant, offers the added benefit of protecting the brain against free radical damage, thus slowing the progression of the underlying disease. Contrast this to the "Band-Aid" approach of conventional medicine that treats only symptoms while possibly worsening the underlying disease. In addition to protecting the brain from free radical damage, glutathione makes the cells more sensitive to dopamine, thereby enhancing the effectiveness of whatever dopamine is produced by the brain. I have found that glutathione therapy works extremely well in combination with L-dopa, allowing patients to use lower doses of the drug.

Yet another benefit of this miraculous therapy is that glutathione treatment dramatically improves the detoxification ability of the liver, which protects the brain against toxins that have been associated with Parkinson's disease.

The only downside of intravenous glutathione therapy is that it is more complicated than simply taking a pill. Glutathione itself can't be taken orally because it is broken down in the stomach before it can reach the bloodstream. There are supplements, such as NAC and alpha lipoic acid, that modestly increase glutathione levels, but if you want to achieve a real glutathione boost, you need to administer it directly into the bloodstream. Researchers around the world are developing oral products designed to raise glutathione, and it is hoped that we will soon have such a product. Until then, intravenous glutathione therapy is the best option.

The body requires a steady source of glutathione to function normally. Since Parkinson's patients are not producing enough of it on their own, they need frequent treatments. Typically, intravenous glutathione therapy is given three times weekly, but some patients may require daily treatments. Glutathione should initially be administered by a qualified health-care professional. Once we know how the patient will respond, and the appropriate dose, we teach a caregiver how to administer the shots at home. It is not difficult and takes only about 20 minutes per treatment. We provide information on where to purchase injectable glutathione, as well as an instructional video as to how to use it.

Due to the overwhelming success of this treatment, there are now well over one thousand physicians in this country using our protocol for glutathione administration. (To locate a physician near you, contact the American Academy for the Advancement in Medicine [ACAM] at 949-598-7666 and see Appendix 4 for more information.) Since glutathione works so well, it begs the question: *Why isn't glutathione therapy part of standard medical treatment for Parkinson's?* I think that there are two reasons. First, glutathione is a natural substance, which means that it cannot be patented. It cannot be owned exclusively by any one pharmaceutical company, and so no one is paying for full-page ads in medical journals touting the benefits of this treatment. In fact, you have to go out of your way to find out about it! Second, conventional medicine is reluctant to consider any treatment that has not withstood the scrutiny of the double-blind, placebo-controlled study conducted in the United States. The very good news is that the FDA has now approved our protocol using glutathione to treat Parkinson's disease in a research study at a major medical center. In our study, we will compare the progress of 10 Parkinson's patients on glutathione therapy to those taking a placebo. We plan on publishing results in the spring of 2005. It is my hope that this study—and this book—will bring this treatment to the attention of both Parkinson's patients and medical professionals alike, enabling other Parkinson's patients to live better quality lives. (For more information, see Appendix 4.)

Nutrition Advice

Parkinson's patients should follow the basic Better Brain meal plan (chapter 5). Since your body is already in toxic overload, it's best to avoid processed foods, which are full of chemicals. Stick to fresh, whole foods. Eat organic, pesticide-free produce. Use free-range, hormone-free, antibiotic-free poultry and meat. Drink filtered water.

Don't exceed two cups of coffee daily.

Constipation is a problem for many Parkinson's patients. Eating enough fiber-rich foods daily (fruits, vegetables, and whole grains) can help. Make sure that you drink plenty of water. If the problem persists, use a natural laxative, such as rice fiber, which is sold at health food stores. Taking a magnesium supplement daily (I recommend 500 mg of magnesium aspartate) can really help.

You should follow the Tier 3 supplements (see chapter 6 and Appendix 1).

Supplement Regimen for Parkinson's Patients

SUPPLEMENT	A.M.	P.M.
DHA	300 mg	300 mg
Co-Q10	100 mg	100 mg
Vitamin E	400 IU	
Vitamin C	500 mg	500 mg
Alpha lipoic acid	200 mg	
N-acetyl-cysteine (NAC)	400 mg	400 mg
Phosphatidylserine	100 mg	50 mg
Acetyl-L-carnitine	400 mg	400 mg
Ginkgo biloba	60 mg	
Vitamin D	400 IU	
Vinpocetine*	5 mg	5 mg
Vitamin B complex supplement:*		
B1 (thiamine)	50 mg	
B3 (niacin as niacinamide)	50 mg	
B6 (pyridoxine)	50 mg	
Folic acid	400 mcg	400 mcg*
B12 (cobalamin)	500 mcg	500 mcg*

*Look for a B-complex supplement containing several B vitamins in one capsule or pill.

*In addition to your B complex in the A.M., take an additional 500 mcg of B12 and 400 mcg of folic acid in the P.M.

*Vinpocetine is required only for people with high homocysteine or a history of vascular dementia or coronary artery disease. It should be avoided by people taking Coumadin (warfarin), a blood thinner.

A Long Time Coming

At the Permutter Health Center, our Parkinson's patients have been using Co-Q10 with great success for over 10 years. The rest of the world is finally catching on to the importance of this critical antioxidant in brain disorder. In a recent study published in the *Archives of Neurology*, Parkinson's patients were given Co-Q10, which is an antioxidant and neuronal energizer. The researchers concluded "Coenzyme Q10 was safe and well tolerated at doses of up to 1200 mg per day. Less disability developed in subjects receiving the highest dosage. Coenzyme Q10 appears to slow the progressive deterioration of function in PD, but these results need to be confirmed in a larger study." While the best results in this particular study were observed at 1,200 milligrams per day, it's important to note that the brain requires a variety of antioxidants working in concert to provide maximum protection. Since our protocol encompasses all the important brain antioxidants, it's not necessary to provide such a large amount of Co-Q10.

16   Multiple Sclerosis (MS)

A PROGRESSIVE DISEASE of the central nervous system, MS affects nearly four hundred thousand people in the United States, about two-thirds of them women. It is typically diagnosed between the ages of 20 and 40, although it can strike younger or older people. It is believed to be an autoimmune disease, although there are still many more questions about this disease than answers. Here's what we know so far. In MS, the body's own immune cells attack a particular part of the nerve cell, the myelin sheath, which is the fatty substance that surrounds and protects the nerve fibers of the brain, optic nerves, and spinal cord. The destruction of myelin (demyelination) leaves hardened scar tissue (sclerosis) along the covering of nerve cells, which disrupts the transmission of nerve impulses in the damaged region. Why the immune system begins to attack myelin is still unknown, although there is good evidence that the immune response may be triggered by a virus or a bacterial infection.

The disease is categorized on the basis of the tempo of its progression. The most common type of MS is called relapsing-remitting MS. Patients may experience periodic flares or episodes, which are then followed by a remission period, which can last for many years. There are more serious types of MS that progress faster, but fortunately these are in the minority. In severe cases, MS can be a crippling, debilitating, and life-threatening disease. The overwhelming majority of people diagnosed with MS, however, live a full life span and can lead fairly normal lives. Most remain mobile, although some may eventually need to use a cane or a walker.

There is no cure for MS, although there are some drugs, like the interferons, that may be helpful, at least initially. Recently, however, the long-term usefulness of these drugs has been questioned. In my experience, the best treatment approach is one that emphasizes nutritional therapy. The right dietary protocol can reduce the need for drugs, slow down the progression of this disease, and substantially prolong remission periods in many people.

This disease can be difficult to diagnose because the symptoms are so varied and can mimic many other neurological or psychiatric disorders, such as stroke or ALS. Common symptoms include:

- Tingling or numbness in the limbs
- Slurred speech
- Loss of vision or blurred vision, often in one eye at a time
- Chronic fatigue
- Bladder or bowel problems
- Depression
- Forgetfulness or difficulty concentrating
- Problems with sexual function
- Difficulty walking and poor balance

An MRI (magnetic resonance imaging) scan of the brain and spinal cord can help to provide a definitive diagnosis. This study can detect the myelin destruction in the central nervous system, which appears as white lesions in the brain and spinal cord. These lesions, however, do not automatically mean that you have MS. For example, white lesions seen on a brain MRI could be a sign of an entirely different problem called gluten sensitivity, which can produce many of the same symptoms as MS. Patients who suspect they have MS, or have been diagnosed with MS, should be tested for gluten sensitivity (discussed later in this chapter).

Getting to the Cause

Why would the body's own immune cells begin to attack the precious myelin covering on nerve cells? It's long been suspected that MS is triggered by an infection that disrupts normal immune function, and at least 16 different viruses and bacteria have been cited as possible triggers for MS. The most convincing evidence to date points to one bacteria in particular, *Chlamydia pneumoniae,* a fairly common bacteria that is the cause of pneumonia, pharyngitis, bronchitis, and several chronic diseases. It may also play a causative role in other diseases characterized by ongoing inflammation, including rheumatoid arthritis and heart disease. In one study conducted at Vanderbilt School of Medicine, researchers found evidence of the bacteria in the spinal fluid of 100 percent of the 37 MS patients in their study. Chlamydia can easily be treated with antibiotics, notably doxycycline and tetracycline. (Doxycycline may be more effective because it more readily enters the brain.) The information linking chlamydia to MS is very new, and there are no official, FDA-approved antibiotic protocols for MS. Nevertheless, many physicians who treat MS—myself included—often prescribe an antibiotic regimen for MS patients. Antibiotics are not a cure for MS, but wiping out a possible bacterial trigger may help restore more normal immune function.

There is another microorganism that could play a role in MS—*Candida albicans,* a type of yeast infection. Several studies have suggested a link between yeast overgrowth in the intestines or gut and autoimmune infections in general.

What's so bad about yeast overgrowth? The gut contains more than four hundred different types of microorganisms, many of which are so-called good bacteria that aid in digestion. There are several different kinds of good bacteria, including *Lactobacillus acidophilus,* best known as the agent used to culture yogurt. An overgrowth of yeast can destroy the good bacteria, which can create digestive disorders, as well as aggravate the immune system and promote inflammation, which can spread from the gut to other parts of the body.

There is an epidemic of yeast overgrowth today, primarily triggered by the high amount of refined sugar in the diet from processed foods (yeast loves sugar). It's also important to note that antibiotics can wipe out friendly as well as unfriendly bacteria, which can pave the way for yeast overgrowth. If you take antibiotics for MS or any other problem, it's important to take supplementary "good bacteria," also known as probiotics, to maintain the normal ecology of your gut.

At the Perlmutter Center, we published a study in which we examined 10 MS patients for the presence of specific antibodies against *Candida albicans* that would suggest the presence of infection. We found elevated levels of a particular immunoglobulin (immune cell that binds to *Candida*) in 7 out of 10 of these MS patients. We then ordered a special test called a comprehensive digestive stool analysis (CDSA) to determine which helpful bacteria may be deficient, so we could design more targeted treatments. In nine of our MS patients, we found an excess of bad bacteria and a deficiency of good bacteria, and in particular, eight of the nine were low in lactobacillus bacteria. I believe that all MS patients should be checked for yeast overgrowth and treated appropriately. I'm not suggesting that this is going to cure MS, and that if you correct the imbalance in the gut, all your health problems will go away. I do believe, however, that reducing a potential irritant to the immune system could help reduce the overactivity of the immune response that is damaging nerve cells.

The ABC of Drug Therapy

There are five FDA-approved drugs to treat MS, including Avonex, Betaseron, and Copaxone, known as the ABC drugs, Novantrone, and Rebif. These drugs are all immunosuppressants that dampen the immune system's attack on the nervous system. They have to be administered by injection and are rather pricey. I'm not a big fan of these drugs because I don't think that they are that effective, and they all have side effects, such as flulike symptoms, headache, fatigue, depression, and blood cell abnormalities. Many patients tell me that they feel terrible on them. In addition, steroids are often prescribed to help control a flare but cannot be used long term because of dangerous side effects, including the thinning of skin and bone, cognitive changes, elevated blood sugar, and yeast overgrowth.

Even after they go into remission, many MS patients feel compelled to stay on these powerful drugs for long periods out of fear that if they stop taking them, they will have an attack. There is simply no clear-cut evidence that this is true. The nature of MS is that it ebbs and wanes, often mysteriously. Whether you are taking drugs or not, it's possible that you will go into remission, or worsen.

At the Perlmutter Center, we have had excellent results with our MS protocol, and this success is now being reported by doctors in many other countries who are utilizing our program. The foundation of our program includes a rigorous diet and supplement regimen targeted specifically to stop the inflammatory process that is destroying myelin. The diet that we recommend for MS is stricter than the Better Brain meal plan in chapter 5, but it is well worth the extra effort. I can count on one hand the number of patients adhering to our program who have required additional drug therapy. In addition, we have seen great results with hyperbaric oxygen therapy, a treatment that is widely used for MS in Europe. To be sure, we don't have a cure for MS, but I believe that we are able to keep our patients feeling better and in remission for long periods at a time without forcing them to use drugs that interfere with their quality of life.

The Swank MS Diet

We are strong advocates of the Swank MS Diet, designed by Dr. Roy Swank, M.D., Ph.D., more than 50 years ago. Dr. Swank was the first researcher to discover a correlation between the consumption of dietary fat—specifically, saturated fat—and the incidence of MS. Dr. Swank observed that the incidence of MS is much higher in northern latitudes, where winters are long and the weather is cold, than in southern latitudes, where summers are long and the weather is typically warm. Researchers have long suspected that exposure to sunshine or vitamin D may have a protective effect. Dr. Swank, however, noticed another major difference between northern and southern latitudes—diet. People in northern latitudes tended to eat a diet higher in animal fat with more meat and dairy than those who lived in southern latitudes, where they ate more fruits and vegetables. On the basis of these findings, Dr. Swank tested a very-low-fat diet on MS patients and discovered that their MS progressed much less rapidly than in patients consuming a diet higher in fat. In a study following 146 patients on his very-low-fat diet for up to 17 years, Dr. Swank noted that, "if treated early in the disease, before significant disability had developed, a high percentage of cases remained unchanged for up to 20 years."

The Swank diet is a low-fat, vegetarian-based diet that allows only a small amount of saturated fat (from meat, eggs, and full-fat dairy products) and emphasizes the good polyunsaturated fats found in vegetables, seeds, nuts, and fish. In fact, the first year you are on the diet, you are not supposed to eat any red meat or milk fat at all, but you are allowed to eat some chicken and fish. Given the inflammatory nature of MS, this makes perfect sense. Saturated fat promotes inflammation throughout the body. Polyunsaturated good fats (omega 3 and omega 6 fatty acids) reduce inflamation and have a soothing effect on the immune system. When your nervous system is "on fire," it makes sense to do whatever you can to calm it down.

For more information on the Swank diet, read Dr. Swank's book, *The Multiple Sclerosis Diet Book: A Low-Fat Diet for the Treatment of MS* (New York: Doubleday, 1987). It provides important information you need to adapt the diet successfully to your lifestyle.

Hyperbaric Oxygen Therapy

In hyperbaric oxygen therapy, patients are exposed to pure oxygen under increased atmospheric pressure. Patients are placed in a clear glass chamber where they breathe 100 percent oxygen under increased pressure. During the treatments, which usually last for 60 to 90 minutes, patients can read, watch a video, or relax. The therapy works by boosting the levels of oxygen within the cells of the body, which promotes healing. It also increases the ability of disease-fighting white blood cells to destroy bacteria, viruses, and other toxic invaders within the body. At the same time, it can help tone down an inappropriately overactive immune system, as in the case of MS. In addition, enhanced tissue oxygenation creates a more favorable environment for myelin repair, especially in conjunction with essential fatty acids and other supplements.

Among my patients, I have found that hyperbaric oxygen therapy can help relieve symptoms of MS and slow down the progression of the disease. In 1982 a groundbreaking article published in the highly respected *New England Journal of Medicine* reported on a one-year, randomized, double-blind, placebo-controlled study of MS patients receiving hyperbaric oxygen therapy versus a group of untreated patients. The results were extremely impressive. Only 12 percent of the treated group had worsening of their symptoms, as compared to 55 percent of the untreated group. Moreover, many of the patients who received the treatment experienced a real improvement in symptoms, including better mobility and bladder control and diminished visual problems. A follow-up study published four years later confirmed the good results seen in the initial study. Furthermore, it noted another benefit experienced by the treated patients: they maintained better physical coordination than the untreated group.

At the Perlmutter Center, we have been treating MS patients using hyperbaric oxygen for the past four years, and I can attest to the profound effectiveness of this therapy. Fortunately, there are now several centers around the country willing to provide this valuable treatment for MS patients. To learn more about hyperbaric oxygen, visit www.perlhealth.com.

Antigliadin Antibody Test

Could a protein found in grain be hurting your brain and nervous system? If you are one of the millions of people who are sensitive to gluten, the answer is definitely yes. Gluten is a protein found in many different grains, including wheat, rye, barley, buckwheat, amaranth, and spelt. An astounding 1 in 250 Americans is "gluten intolerant," and suffers from a condition known as celiac disease. For these people, eating gluten in any form can cause severe health problems, including chronic upset stomach, malabsorption of nutrients, weight loss, and bone pain. *Gluten can also cause significant neurological problems, including telltale white lesions in the brain that are similar to the kind of damage seen in patients with multiple sclerosis, as well as memory loss and confusion.* I have had numerous patients come to me with cognitive or nerve problems that were caused or aggravated by gluten intolerance. The severity of the symptoms can be almost unbelievable. For example, I recently treated a woman who came to me after she had been diagnosed with MS. This patient had complained of a sensation of weakness and numbness down her right arm, a sign of nerve damage; GI distress; and confusion. An MRI scan of her brain revealed inflammatory lesions of the kind associated with MS. The doctor who had diagnosed her with MS had also prescribed medication that was making her feel even more sick to her stomach. Given her symptoms, there was reason to suspect MS, but I was aware of studies that had shown that gluten intolerance could produce the same brain lesions and neurological symptoms. There have been several recent reports in neurology journals on cases such as this in which MS symptoms were actually due to a problem with gluten. Sure enough, my patient tested positive for antigliadin antibodies and within a few weeks of following a gluten-free diet, her symptoms vanished. If you are experiencing any neurological symptoms that suggest MS, I urgently recommend the gluten test. It's a simple blood test called the antigliadin antibody test, and it detects antigliadin antibodies in the blood.

You may wonder how such common foods as grains can cause serious, even life-threatening, symptoms. When people who are gluten intolerant eat foods containing gluten, their immune cells target this protein as they would a bacteria or virus. This creates an inflammatory response that eventually destroys the villi, the tiny, fingerlike protrusions in the lining of the intestine through which all our nutrients are absorbed. In addition, this inflammatory response can directly affect brain tissue, much as is seen in MS. Over time, gluten intolerance can lead to severe malnourishment and nutrient deficiencies. In fact, B12 deficiency, which is linked to many neurological problems, is very common in people with this problem. Nutrient deficiency by itself can lead to elevated homocysteine and weakened antioxidant defenses, which is serious, but which becomes even more toxic when combined with chronic inflammation. Every part of the body, from your bones to your brain, can be affected. People with untreated gluten intolerance are not only prone to develop neurological symptoms but are at higher risk of osteoporosis, rheumatoid arthritis, lymphoma (a cancer of blood cells or the lymphatic tissue), and other autoimmune problems.

It can be difficult to diagnose gluten intolerance because the symptoms can be so diverse, and seemingly unrelated to a gut problem. In addition to the more obvious abdominal bloating and pain, symptoms can also include unexplained anemia (due to poor iron and/or B12 absorption), joint pain, skin rashes, weight loss, behavioral changes, and the neurological problems noted earlier. Patients can go for years without getting a proper diagnosis and may be treated for other conditions along the way that would vanish if gluten was eliminated from their diet.

Gluten intolerance is effectively treated by eliminating gluten from your diet. Unfortunately, this sounds easier than it is. Gluten is not only found in grains but is also a common ingredient in processed foods; condiments, including soy sauce; and even prescription and over-the-counter drugs. (Experts disagree as to whether gluten is present in oats, but to be on the safe side, I recommend that you steer clear of oats also.) If you are gluten intolerant, you must become an educated food consumer, and that means reading labels and calling manufacturers if you are not certain whether or not a food contains gluten. Many food manufacturers have become more sensitive to this problem, and there are gluten-free versions of many commonly used foods. If you test positive for gluten intolerance, it's essential that you work with a competent nutritionist who can help you design the right eating plan. In order to keep your symptoms at bay, you have to completely eliminate gluten from your diet. You cannot start eating gluten again when your symptoms disappear, or you will regress back to ill health.

Although gluten intolerance can cause chronic digestive distress such as gas, bloating, diarrhea, and general discomfort, sometimes there are no GI symptoms. The only symptoms may be neurological. In fact, according to recommendations made at the 2002 Annual Scientific Meeting of the American Neurological Association, patients with neurological symptoms of unknown origin should be tested for gluten intolerance even in the absence of telltale GI symptoms. I would go even further. Given the ubiquitous nature of gluten in the food supply, I think everybody should ask their doctor to perform this test so that they can alter their diet before they develop symptoms.

I've encountered some savvy patients who have eliminated gluten on their own on the suspicion that they are gluten intolerant. Given the fact that following a gluten-free diet can be difficult, I don't recommend that you do it unless you test positive for the antigliadin antibody. You may be putting yourself through a lot of extra work for no added benefit. If you have already eliminated gluten from your diet because you suspect you have a problem, your test results will not be accurate. Remember, the test looks for the presence of antigliadin antibodies, but if you have not been eating gluten recently, your body would not be producing the antibodies. (For more information on gluten intolerance, see page 286.)

Supplements

For MS patients, we use the Tier 3 supplement regimen (see chapter 6 and appendix 1). In addition, we recommend the following.

Borage oil. This is a great source of GLA, an essential fatty acid that can help reduce brain inflammation. Several studies have shown that MS patients are deficient in essential fatty acids and benefit from supplementation. Take a dose that supplies about 300 milligrams of GLA daily. You can buy borage oil at health food stores and pharmacies.

B12 shots. Vitamin B12 is important in the formation and maintenance of myelin. Many patients find vitamin B12 injections to be very helpful and are taught how to administer the shots themselves. The initial dose is 1 cc (1000 mcg) daily for the first five days, then twice weekly thereafter.

Antibiotic therapy. Because of the possibility that the bacterium *Chlamydia pneumoniae* may be involved in triggering MS, we routinely provide a course of doxycycline (100 mg for 21 days) at the beginning of our protocol, along with the appropriate probiotic (*Lactobacillus acidophilus*) supplement. Interestingly, many of our patients have reported significant symptomatic improvements from this intervention alone.

Follow the Tier 3 supplements (see chapter 6 and appendix 1).

Supplement Regimen for MS Patients		
SUPPLEMENT	A.M.	P.M.
GLA (from Borage oil)	300 mg.	
DHA	300 mg	300 mg
Co-Q10	100 mg	100 mg
Vitamin E	400 IU	
Vitamin C	500 mg	500 mg
Alpha lipoic acid	200 mg	
N-acetyl cysteine (NAC)	400 mg	400 mg
Phosphatidylserine	100 mg	100 mg
Acetyl-L-carnitine	400 mg	400 mg
Ginkgo biloba	60 mg	
Vitamin D	400 IU	
Vinpocetine*	5 mg	5 mg
Vitamin B-complex supplement:**		
B1 (thiamine)	50 mg	
B3 (niacin as niacinamide)	50 mg	

B6 (pyridoxine)	50 mg	
Folic acid	400 mcg	400 mcg***
B12 (cobalamin)	500 mcg	500 mcg***

*Vinpocetine is required only for people with high homocysteine or a history of vascular dementia or coronary artery disease. It should be avoided by people taking Coumadin (warfarin), a blood thinner.

**Look for a B-complex supplement containing several B vitamins in one capsule or pill

***In addition to your B-complex in the A.M., take an additional 500 mcg of B12 and 400 mcg folic acid in the P.M.

Amyotrophic Lateral Sclerosis (ALS)

ALSO KNOWN AS Lou Gehrig's disease, in honor of one of baseball's greatest players who was stricken with it in 1939, ALS is a progressive, degenerative disorder of the part of the nervous system that controls voluntary muscle movement. There is no cure, and there are no conventional treatments to significantly slow down the disease. About 30,000 Americans have this disease, and about 5,600 are diagnosed with it each year. The average age of onset is between 55 and 75. Although it is often described as a "terminal" disease, that word has to be put in perspective. About half of all people with ALS live three or more years after diagnosis; 20 percent go on to live five or more years, and 10 percent survive more than 10 years. I believe that with proper and aggressive treatment, we are able to slow down the progression of this disease and my patients do better than they otherwise would. We have not yet cured a single case of ALS, but we have made progress, and we can offer some hope. Our patients don't decline as rapidly, and even several years after beginning treatment, many show minimal decline.

The degree of severity of ALS varies widely from patient to patient. A minority of patients (between 10 and 20 percent) deteriorate very rapidly no matter what you do, but most do amazingly well on the protocol we use at the Perlmutter Center. The earlier we begin treatment, the better the result.

This disease is often difficult to diagnose initially because of the wide array of symptoms with which it may present. Early in the disease, a patient may experience a weakness of an arm or leg, which could be thought to be due to an overuse injury, a ruptured disc in the spine, or countless other problems. Other symptoms include difficulty with speech or trying to project your voice, chronic fatigue, stiffness of an extremity, twitching of muscles, or problems swallowing. As the disease progresses, patients may have difficulty breathing, which often leads to pneumonia.

There is no one diagnostic test for ALS. A physician must evaluate clinical symptoms, standard laboratory tests, and perform other tests, often including an electromyography (EMG), to check for nerve function, or an MRI of the brain and spinal cord to rule out other possible diseases. Other neurological ailments, such as MS, Parkinson's disease, stroke, nerve injury, disc disease, brain tumor, and even depression must be ruled out. Some of these symptoms could even be caused by a deficiency in vitamin B12. Once other ailments are excluded, if the neurological symptoms worsen, ALS is then considered as a diagnosis. In light of the implication of the diagnosis, I recommend that patients always seek a second opinion from a major medical center. The cause of ALS is very much a mystery. About 5 percent of all cases are genetic, but the overwhelming majority of cases appear to strike at random. Some occupations put you at greater risk than others. There is a higher incidence of ALS among pilots, electrical workers, and veterans of the Gulf War of 1991, which indicates that exposure to environmental toxins may trigger this disease in some people. Moreover, there may be an elevated risk of ALS among high-performance athletes. Recent studies also suggest that an unidentified infection may be involved in this disease: researchers were able to significantly delay the onset and slow the progression of symptoms in mice that were genetically disposed to develop ALS by treating them with a common antibiotic, minocycline. Although there is no clearcut explanation as to why the antibiotic worked, researchers theorized that it may have eliminated an unidentified infection. It is also known, however, that minocyline may block the release of an enzyme that kills mitochondria, the energy-producing part of the cell. The successful results in laboratory animals has prompted the National Institutes of Health (NIH) to sponsor human clinical trials evaluating minocycline therapy in ALS patients at the University of New Mexico.

We may not know how ALS picks its victims, but we do know the mechanism by which it does its dirty work. The disease is characterized by a marked deficiency in cellular energy production by the mitochondria in nerve cells, which results in excess free radical production. In a sense, ALS is a wildly accelerated and more destructive version of the aging process. Inflammation may also play a role in promoting the destruction of nerve cells. Patients typically have increased levels of the COX-2 enzyme, which greatly enhances inflammation, in their nervous system. This finding has led researchers to conduct clinical studies investigating the effectiveness of common COX-2-inhibiting agents like Celebrex (celecoxib) as a treatment for ALS.

ALS—The Athlete Connection

High-intensity athletes are at an increased risk of developing early ALS. The question has never been fully studied, but I suspect that athletes who are constantly taxing their bodies may be overstressing their energy-producing mitochondria. Ultimately, the mitochondria may wear out, which causes the energy deficiency later in life. Moreover, the harder you work your body, the more free radicals you generate. Parents of endurance athletes should be vigilant in making sure that their children do not overtrain or continually push themselves to their physical limits. It's wise for endurance athletes of all ages to fortify themselves with an adequate amount of antioxidants, especially the neuronal energizers such as Co-Q10 and alpha lipoic acid.

Conventional Treatments

Rilutek (riluzole) is the only drug approved for use for ALS. In the best-case scenario, it may extend the life span of patients by three months—but at a price. Rilutek can cause nasty side effects, including worsening of ALS symptoms, such as fatigue, muscle weakness, and muscle spasticity. It is also expensive, costing up to $7,000 a year. In 1999 the American Academy of Neurology published a monograph outlining the current accepted approach for treatment for ALS. In this report, the group noted that "it is often said that the benefits of riluzole are marginal, but that the side effects are major. . . . The price of the drug is also often criticized, especially if the patient's health care system does not pay for the drug." Personally, I can't see much benefit to using this drug.

The Perlmutter Center's protocol for treating ALS focuses on two primary areas: (1) protecting against free radical damage through glutathione IV treatment, and (2) using supplements to boost cellular energy production and enhance antioxidant coverage.

We also prescribe antibiotic therapy to eliminate any undetected infection and Celebrex to control inflammation.

Intravenous Glutathione Therapy

Since free radical oxidative injury is the ultimate cause of nerve damage in ALS, protecting neurons with potent antioxidant coverage and enhancing energy production to reduce free radical formation is the ultimate goal of treatment. All our ALS patients follow the Tier 3 supplement regimen. We also include intravenous glutathione in our ALS protocol.

Glutathione is perhaps the most effective and beneficial antioxidant in the nervous system and has the added benefit of enhancing mitochondrial energy production.

We generally begin treatment as described in the Parkinson's glutathione protocol at 1,400 milligrams three times weekly. Interestingly, the University of Kansas has adopted our glutathione protocol for research evaluation in the treatment of ALS. We routinely recheck the urine lipid peroxide level every two to three months, even if it is normal. If the urine lipid peroxide level remains abnormally high, we then utilize the Tier 3 Plus program (see page 176). If the lipid peroxide levels still remain abnormally elevated, we then increase the IV glutathione dosage to 2,000 milligrams every other day.

Two New Experimental ALS Treatments

Research in animal models of ALS supports the idea that the antibiotic minocycline and the anti-inflammatory Celebrex could prove to be useful, and there are ongoing clinical trials attempting to demonstrate their effectiveness in human ALS patients. My concern is that these studies may take years to be completed and published, precious time lost for ALS patients and their families. To me, the experimental research available thus far clearly indicates that these approaches stand a good chance of being effective. Furthermore, these are fairly benign interventions that do not put patients at undue risk. When it comes to ALS, time is clearly of the essence, so we choose to move ahead and treat.

We recommend that our ALS patients take the following prescription medications:

- *Minocycline:* 100 mg twice daily, along with a probiotic supplement (*Lactobacillus acidophilus*)
- *Celebrex:* 200 mg twice daily for one month, increased to 400 mg twice a month thereafter

Review these medications with your physician to discuss the potential side effects, as well as possible interactions with other medications that you may be taking.

Supplements

You should follow the Tier 3 supplements regimen (see chapter 6 and Appendix 1). In addition, I recommend adding *creatine monohydrate* to your supplement regimen. This nutritional supplement, which is commonly used by athletes to build muscle mass, plays a pivotal role in the process by which muscle cells use energy. In animal studies, oral administration of ALS extended life span by up to 18 percent. We recommend that our ALS patients take 5 grams of creatine monohydrate daily. Creatine is inexpensive and is available without a prescription at most health food stores. It is available as a powder, which can be mixed with water or juice (1 tablespoon of powder is equal to 1 gram).

Supplement Regimen for ALS Patients		
SUPPLEMENT	A.M.	P.M.
Creatine monohydrate	5 grams (1 tablespoon)	
DHA	300 mg	300 mg
Co-Q10	100 mg	100 mg
Vitamin E	400 IU	
Vitamin C	500 mg	500 mg
Alpha lipoic acid	200 mg	
N-acetyl-cysteine (NAC)	400 mg	400 mg
Phosphatidylserine	100 mg	100 mg
Acetyl-L-carnitine	400 mg	400 mg
Ginkgo biloba	60 mg	
Vitamin D	400 IU	
Vinpocetine*	5 mg	5 mg
Vitamin B-complex supplement:*		
B1 (thiamine)	50 mg	
B3 (niacin as niacinamide)	50 mg	
B6 (pyridoxine)	50 mg	

| Folic acid | 400 mcg | 400 mcg*** |
| B12 (cobalamin) | 500 mcg | 500 mcg*** |

*Vinpocetine is required only for people with high homocysteine or a history of vascular dementia or coronary artery disease. It should be avoided by people taking Coumadin (warfarin), a blood thinner.

**Look for a B-complex supplement containing several B vitamins in one capsule or pill.

***In addition to your B complex in the A.M., take an additional 500 mcg of B12 and 400 mcg folic acid in the P.M.

Supplement Regimens: The Three Tiers

Tier 1: Prevention and Maintenance		
SUPPLEMENT	**A.M.**	**P.M.**
DHA	300 mg	
Co-Q10	30 mg	
Vitamin E	200 IU (d-alpha, *not* dl-alpha form)	
Vitamin C	200 mg	
Vitamin B complex supplement:*		
B1 (thiamine)	50 mg	
B3 (niacin as niacinamide)	50 mg	
B6 (pyridoxine)	50 mg	
Folic acid	400 mcg	
B12 (cobalamin)	500 mcg	
Look for a B-complex supplement containing several B vitamins in one capsule or pill.		

Tier 2: Prevention, Repair, and Enhancement

SUPPLEMENT	A.M.	P.M.
DHA	300 mg	
Co-Q10	60 mg	
Vitamin E	200 IU	
Vitamin C	200 mg	200 mg
Alpha lipoic acid	80 mg	
N-acetyl-cysteine (NAC)	400 mg	
Acetyl-L-carnitine	400 mg	
Phosphatidylserine	100 mg	
Vinpocetine*	5 mg	5 mg
Vitamin B-complex supplement:**		
B1 (thiamine)	50 mg	
B3 (niacin as niacinamide)	50 mg	
B6 (pyridoxine)	50 mg	
Folic acid	400 mcg	400 mcg***
B12 (cobalamin)	500 mcg	500 mcg***

*Vinpocetine is required only for people with high homocysteine or a history of vascular dementia or coronary artery disease. It should be avoided by people taking Coumadin (warfarin), a blood thinner.

**Look for a B-complex supplement containing several B vitamins in one capsule or pill.

***In addition to your B complex in the A.M., take an additional 500 mcg of B12 and 400 mcg of folic acid in the P.M.

Tier 3: Recovery and Enhancement

SUPPLEMENT	A.M.	P.M.
DHA	300 mg	300 mg
Co-Q10	100 mg	100 mg
Vitamin E	400 IU	
Vitamin C	200 mg	200 mg
Alpha lipoic acid	200 mg	
N-acetyl-cysteine (NAC)	400 mg	400 mg
Acetyl-L-carnitine	400 mg	400 mg

Phosphatidylserine	100 mg	100 mg
Ginkgo biloba	60 mg	
Vitamin D	400 IU	
Vinpocetine*	5 mg	5 mg
Vitamin B-complex supplement:**		
B1 (thiamine)	50 mg	
B3 (niacin as niacinamide)	50 mg	
B6 (pyridoxine)	50 mg	
Folic acid	400 mcg	400 mcg***
B12 (cobalamin)	500 mcg	500 mcg***

*Vinpocetine is required only for people with high homocysteine or a history of vascular dementia or coronary artery disease. It should be avoided by people taking Coumadin, a blood thinner.

**Look for a B-complex supplement containing several B vitamins in one capsule or pill.

***In addition to your B complex in the A.M., take an additional 500 mcg of B12 and 400 mcg of folic acid in the P.M.

Recipes

THESE MENUS AND RECIPES were designed by J. Gabrielle Rabner, M.S., R.D., a holistic nutritionist who is based in Naples, Florida, and Montclair, New Jersey.

Chicken with Honey Mustard Sauce

½ cup honey
1 tablespoon apple cider vinegar
2 tablespoons Dijon mustard
1 3–4 lb. chicken cut into eighths

1. Preheat oven to 400°F.
2. To make basting sauce, combine all ingredients except chicken pieces.
3. Place chicken pieces in large roasting pan.
4. Bake chicken for 30 minutes.
5. Reduce oven temperature to 375°F. Spread sauce, using a basting brush, over both sides of chicken and continue to bake, basting often, for an additional 30 minutes.
6. Bake until crisp and juices of chicken run clear.

Haddock Stew

2 leeks, washed well and cut into thin slices

1 tablespoon extra-virgin olive oil

4 small turnips, cut into chunks

1 sweet potato, cut into chunks

2 carrots, sliced

2 stalks celery, sliced

½ cup sliced shiitake mushrooms

2 tablespoons barley miso paste dissolved in 3 cups filtered or bottled hot water

1 tablespoon dried basil

fresh parsley sprigs

Celtic salt to taste

¾ lb. haddock, cut into 1-inch pieces

1. In a soup pot, sauté leeks in olive oil until tender.
2. Add turnips, sweet potato, carrots, celery, and mushrooms and cook, stirring, about 5 minutes.
3. Add barley miso stock and basil, cover, and bring to a boil. Reduce heat to simmer and cook about 10 minutes or until vegetables are tender.
4. Add haddock and simmer uncovered until fish is cooked through and stew is thickened, about 10 minutes.
5. Season with salt to taste and serve topped with parsley sprigs.

Baked Garlic and Soy Sauce Tofu

1 block tofu, rinsed, dried, and cut into ½-inch slices

¼ cup tamari soy sauce

6 garlic cloves, crushed

1 tablespoon olive oil

1. Preheat oven to 400°F.
2. Brush baking pan to hold tofu slices with 1 teaspoon olive oil.
3. Combine soy sauce, garlic, and remaining 2 teaspoons olive oil in a shallow bowl.

4. Dip tofu slices into mixture to coat and place, not overlapping, in the baking pan.
5. Bake in oven for about 30 minutes or until nicely browned.

Gingered Flank Steak

1 flank steak
4-inch piece fresh ginger root, peeled and grated
3 tablespoons tamari soy sauce
3 chopped garlic cloves
2 tablespoons extra-virgin olive oil

1. Combine all ingredients except steak.
2. Place steak in a large baking or roasting pan and pour sauce over steak, turning steak to cover both sides. Marinate for about 1 hour.
3. Broil steak or barbecue to desired doneness.

Basic Flaxseed and Olive Oil Dressing with Variations

½ cup extra-virgin olive oil
¼ cup flaxseed oil
2 tablespoons fresh lemon juice
½ teaspoon prepared mustard
6 walnut halves (optional)

Combine all ingredients in a blender and blend until smooth, about 1–2 minutes.

Variations

FOR OLIVE AND PUMPKIN SEED DRESSING, substitute pumpkin seed oil for the flaxseed oil and blend all ingredients as directed.

FOR WALNUT AND OLIVE DRESSING, substitute walnut oil for the flaxseed oil and blend all ingredients as directed.

FOR LEMON, PARSLEY, AND FLAXSEED OIL DRESSING, use parsley instead of the mustard and proceed as directed.

Whey or Rice Protein Shake and Variations

2 scoops rice or whey protein powder
2 tablespoons flaxseeds, ground in a blender or coffee grinder
1 cup fresh fruit, berries, or precut large fruit, frozen
1 cup filtered or bottled water or yogurt or liquid of choice

Combine all ingredients in a blender and blend until frothy and thickened.

Shake Suggestions

FOR A RICE PROTEIN SHAKE WITH BANANA, STRAWBERRIES, AND GROUND FLAXSEEDS, use ½ cup strawberries and a medium banana. Combine and blend as directed.

FOR A RICE PROTEIN SHAKE WITH PINEAPPLE, APRICOTS, AND GROUND FLAXSEEDS, use ½ cup fresh pineapple and 2 apricots. Combine and blend as directed.

FOR A BANANA, PAPAYA, AND GROUND FLAXSEED SHAKE, use banana and papaya in place of the fruit in the basic recipe and prepare as directed.

FOR A CARROT AND SPINACH JUICE SHAKE WITH GROUND FLAXSEEDS AND RICE PROTEIN: in a juicer, juice 3 carrots and 1 bunch of spinach. Add ½ cup filtered or bottled water, ground flaxseeds, and rice protein as directed.

FOR THE WHEY PROTEIN SHAKE WITH PINEAPPLE, STRAWBERRY, AND GROUND FLAXSEEDS: use ½ cup of each fruit and prepare as directed.

Steamed Purslane

1 cup uncooked purslane per person
filtered or bottled water
Celtic salt to taste
olive oil (optional)

1. Rinse purslane and trim any leggy stems.
2. In a saucepot, place ½ inch filtered or bottled water. Place steamer basket in pot and put purslane into basket.
3. Cover pot and bring water to a boil. Steam on low heat for a few minutes or until purslane is tender. Serve immediately, plain or with a drizzle of olive oil and salt to taste.

Stir-Fried Vegetables and Marinated Tempeh

8 ounces tempeh, cut into 1-inch cubes
1 tablespoon tamari soy sauce
¼ cup mirin*
2 tablespoons fresh lemon juice
1 tablespoon raw honey
3 cloves garlic, crushed
2-inch piece fresh ginger root, grated
    (or use ¼ teaspoon powdered ginger)
2 tablespoons extra-virgin olive oil

½ each green and yellow pepper, cut into 1-inch pieces
1 onion, sliced thin
1 green zucchini, sliced
1 carrot, sliced
1 stalk celery, sliced
1 cup sliced red cabbage

*Japanese rice wine, found in health food stores.

1. Combine soy sauce, mirin, lemon juice, honey, garlic, ginger and 1 tablespoon olive oil. Marinate tempeh pieces in marinade for about 1 hour.
2. Meanwhile prepare vegetables and stir-fry in the remaining 1 tablespoon olive oil until crisp-tender.
3. Broil tempeh about 5 minutes, then add to stir-fried vegetables. Continue to stir-fry mixture for about 5 minutes longer.
4. Any desired vegetables can be substituted and stir-fried.

Vegetarian Caesar Salad

1 head romaine lettuce, torn into bite-sized pieces

Dressing
1 whole piece of kelp, cut with a scissors into ¼-inch pieces
juice of two lemons
1 cup extra-virgin olive oil
5 cloves garlic
¼ cup Parmesan cheese, grated

Croutons
4 slices whole wheat bread, toasted
¼ cup olive oil
2 cloves garlic
Celtic salt to taste

1. Preheat oven to 450°F.
2. Make the dressing by blending in a blender or food processor: olive oil, lemon juice, kelp, garlic, and cheese. It may be thinned with a little water if needed.
3. To make the croutons, cube the bread. Then blend the olive oil, garlic, and salt to taste. Toss the cubes in the oil mixture. Place cubes on a cookie sheet in a single layer and bake in oven for about 4 minutes on each side.
4. In a large salad bowl, toss the romaine with the dressing. Add the croutons and serve.

Whole-Grain Millet Drizzled with Flaxseed Oil and Topped with Sliced Almonds

1 cup millet grain, soaked 6 hours or overnight in filtered or bottled
   water to cover

2 cups filtered or bottled water

2 tablespoons flaxseed oil

2 tablespoons sliced almonds (may be purchased already sliced)

1. Drain soaking water from millet. Bring 2 cups of water to the boil.
   Add millet and simmer, covered, 30–40 minutes or until tender.
2. Mix flaxseed oil and almonds into millet and serve.

Variation

FOR POACHED EGG ON MILLET, prepare basic millet as directed.

TO MAKE POACHED EGGS, bring 1–2 inches of water and 1 teaspoon of
white vinegar in a small skillet to a simmer. Drop eggs in one at a time
and cover for 3–5 minutes or until whites are firm. Place on top of
cooked millet.

Oven-Roasted Vegetables

1 cup each carrots, sweet potatoes, mushrooms, leeks, beets, turnips, sliced
   or cut into chunks

1. Preheat oven to 425°F.
2. Brush vegetables with extra-virgin olive oil.
3. Bake vegetables in a roasting pan or baking dish in oven for about 40
   minutes or until vegetables are crisp-tender.

Garlic Hummus with Cut-up Vegetables

1 15-oz. can organic chickpeas (garbanzo beans) drained, reserving liquid*

3 large cloves garlic

2 tablespoons tahini (sesame paste)

juice of 1 lemon

fresh parsley sprigs (optional)

assorted cut-up vegetables of choice, such as baby carrots, sliced cucumber, celery sticks, etc.

1. Place all ingredients except parsley and cut-up vegetables into food processor or blender. Add 2 tablespoons reserved drained liquid from beans.
2. Blend until a smooth paste forms, adding more liquid if necessary.
3. Garnish with parsley sprigs and serve with assorted cut-up vegetables.

*Eden organic beans is a good choice, low in sodium and with a good crisp texture.

Buckwheat Pancakes with Maple Syrup and Fresh Berries

buckwheat pancake mix*

½ to 1 cup assorted fresh berries such as blueberries, blackberries, raspberries

maple syrup to taste

1. Prepare pancake mix per package directions.
2. Mix the berries and place in serving bowl.
3. Serve the pancakes topped with maple syrup and berries.

*Arrowhead Mills is a good organic brand.

Variations

FOR BUCKWHEAT AND FLAXSEED PANCAKES, add ¼ cup flaxseeds to the pancake mix before making pancakes. Can be served with or without berries.

Guacamole Dip with Walnuts

2 avocados
2 tablespoons fresh lemon juice
1 or 2 cloves garlic, crushed
Celtic salt to taste
½ cup chopped walnuts

1. Scoop out avocados into bowl and mash with lemon juice.
2. Add remaining ingredients and mix well.
3. Serve with cut-up raw vegetables or corn chips.

Baked Chicken with Turnips and Parsnips

1 chicken, cut up into eighths
1 onion, sliced
2 teaspoons extra-virgin olive oil
2 medium-sized turnips, sliced
2 medium-sized parsnips, sliced
garlic powder to taste
Celtic salt to taste

1. Preheat oven to 400°F.
2. Place chicken pieces in roasting pan. Rub with 1 teaspoon olive oil.
3. Sprinkle pieces with garlic powder and Celtic salt.
4. Rub 1 teaspoon of olive oil on the turnip and parsnip slices. Place them in pan around the chicken.
5. Bake chicken and turnip and parsnip slices for 50 to 60 minutes or until juices of chicken run clear.

Spinach Mushroom Frittata

5 large organic eggs, beaten with 1 tablespoon filtered or bottled water
2 tablespoons extra-virgin olive oil

½ medium onion, sliced thinly

2 cups baby spinach

1 cup mushrooms of choice, sliced

¼ cup Monterey Jack raw milk cheese, grated

Celtic salt to taste

1. Preheat oven to 350°F.
2. In a large ovenproof fry pan, heat 1 tablespoon olive oil and cook onions and mushrooms on medium low heat until soft. Add spinach and blend mixture lightly.
3. In a large bowl, add cheese to egg and water mixture. Stir in cooked vegetables.
4. Heat remaining tablespoon of olive oil in fry pan. Add egg-vegetable mixture and cook on low until cooked halfway through.
5. Place pan with frittata in oven, uncovered, and bake until nicely browned and puffy. Salt to taste.

String Beans with Pumpkin Seeds and Garlic Vinaigrette

1 cup string beans per person

2 tablespoons pumpkin seeds per person

garlic vinaigrette (see recipe that follows)

1. Cook string beans until crisp-tender.
2. Top with pumpkin seeds and garlic vinaigrette to taste.
3. Serve immediately.

Basic Garlic Vinaigrette

1 cup extra-virgin olive oil

¼ cup vinegar of choice, such as apple cider or balsamic

½ teaspoon Dijon mustard

Celtic salt to taste

2 teaspoons minced fresh garlic

1. Mix vinegar, salt, and mustard using wire whisk or blender or food processor.
2. Add oil in slow, steady stream as you are whisking or blending.

Baked Apples with Walnuts and Figs

4 large apples such as Granny Smith or Gala

2 teaspoons ground cinnamon

3 tablespoons maple syrup

2 dried figs, cut up (use kitchen scissors) in small pieces

2 tablespoons walnuts, chopped fine

1 cup filtered or bottled water

1. Preheat oven to 350°F.
2. Core apples and place in baking dish.
3. Mix cinnamon, maple syrup, figs, and walnuts. Divide mixture among apple cavities. Any remaining mixture can be poured over apples.
4. Pour water around apples and bake in oven, uncovered, for 1 hour or until apples are soft.
5. Serve hot or cold.

Scrambled Tofu with Onions and Broccoli

1 block firm tofu, washed and dried

1 onion, sliced

1 cup broccoli florets, cut into small pieces

1 tablespoon extra-virgin olive oil

1. Heat olive oil on medium-low heat in fry pan.
2. Sauté onions and broccoli about 3 minutes.

3. Cut tofu into chunks and add to fry pan. Using a fork, continue to mash tofu to incorporate it into vegetables until desired degree of doneness, about 5 minutes.

Waldorf Salad and Homemade Mayonnaise

4 medium apples, chopped (must be peeled if not organic)
2 celery stalks, sliced thinly
½ cup homemade mayonnaise (see recipe that follows)
lettuce leaves

1. Toss apples and celery in enough mayonnaise to make the mixture slightly creamy.
2. Serve over lettuce leaves.

Homemade Mayonnaise

1 egg
½ teaspoon dry mustard
2 tablespoons fresh lemon juice
1 cup extra-virgin olive oil
Celtic salt to taste

1. In a blender or food processor, combine egg, mustard, salt, lemon juice, and ¼ cup oil.
2. With the machine running, add the remaining oil in a slow, steady stream until mixture thickens. If too thick, add a little filtered or bottled warm water.
3. Can be stored in the refrigerator for a week.

Brown Rice Pudding with Apple Slices, Raisins, and Pumpkin Seeds

1 cup brown rice, cooked

1 apple, sliced (must be peeled if not organic)

2 tablespoons filtered or bottled water or milk

2 tablespoons raisins

2 tablespoons pumpkin seeds

1. Warm the apple slices in the brown rice over low heat (add water or milk).
2. Add raisins and pumpkin seeds and serve.

French Onion Soup

6 medium onions, thinly sliced

3 tablespoons extra-virgin olive oil

4 bouillon cubes dissolved in 4 cups filtered or bottled water or miso stock (recipe follows)

1 white potato, chopped

Celtic salt to taste

¼ cup white wine (optional)

toasted Ezekiel sprouted grain bread, cut into quarters

grated Parmesan and Gruyère raw milk cheese to taste

Miso Stock

¼ cup barley miso paste

4 cups filtered or bottled water

1. If using miso stock, bring 4 cups water to a boil. Allow to cool for 3 minutes. Add miso paste and stir to dissolve thoroughly. Put aside.
2. In a large soup pot, sauté onions in olive oil on low heat, stirring, until transparent.

3. Add stock or bouillon water and potato and cook, covered, on low heat until onions are tender, about 30 minutes.
4. Stir in wine, if using, and Celtic salt to taste. Simmer 30 minutes. Use a wire whisk to distribute potato evenly.
5. Pour soup into ovenproof serving bowls and float bread squares on soup.
6. Sprinkle with grated cheeses to taste.
7. Broil for about 3 minutes or until cheese melts.
8. Serve at once.

Cantaloupe and Tamari Pumpkin Seeds

cantaloupe cut into quarters, seeds removed
½ cup raw pumpkin seeds
2 tablespoons tamari soy sauce

1. In a large skillet, heat pumpkin seeds over medium-low heat until almost popping. Stir constantly and shake pan so seeds do not burn.
2. Remove from heat and add tamari soy sauce. Toss seeds to mix thoroughly.
3. Remove seeds from pan to cool.
4. Distribute seeds among the cantaloupe quarters and serve.

Salade Niçoise on a Bed of Greens

Salad Dressing
1 tablespoon wine vinegar
3 tablespoons extra-virgin olive oil
3 tablespoons walnut oil
1 teaspoon Dijon mustard
1 clove garlic, crushed
Celtic salt to taste

Blend all ingredients well, place in screw-top jar, and set aside.

Salad

1 pound string beans, cut into ½-inch lengths and steamed 3 minutes

1 green pepper, sliced thinly

2 celery stalks, sliced thinly

2 medium potatoes, cooked, sliced, and cooled

10 ripe olives

1 red onion, sliced thinly

2 tablespoons finely chopped parsley

2 tablespoons finely chopped green onion

3 hard-cooked eggs, quartered

1 cup cherry tomatoes

mesclun greens

1. In a large bowl, combine the first eight ingredients. Mix well. Ingredients may at this point be refrigerated until serving time.
2. When ready to serve, combine dressing and mixed ingredients.
3. Place salad in serving bowl.
4. Garnish with eggs and tomatoes and serve over mesclun greens.

Black Bean Dip

1 15-oz. can black beans, drained*

½ cup salsa

juice of ½ lemon

½ cup chopped onion

¼ cup chopped green pepper

2 cloves garlic, minced

assorted vegetables

1. Blend all but vegetables in a blender or food processor.
2. Serve with cut-up vegetables or crackers.

*Eden low sodium is recommended.

Baked Tilapia

1 pound tilapia
1 leek, washed well and thinly sliced
2–3 cloves garlic, minced
garlic powder to taste
juice of ½ lemon

1. Preheat oven to 350°F.
2. Line a baking pan with leeks and garlic.
3. Place tilapia fillets on top of vegetables.
4. Sprinkle garlic powder on fish and top with fresh lemon.
5. Bake in oven for about 20 minutes or until fish flakes easily.

Stuffed Mushrooms

4 Portobello mushrooms
½ cup minced onions
1 teaspoon minced garlic
2 tablespoons extra-virgin olive oil
2 tablespoons fresh minced parsley
½ cup whole wheat bread crumbs*
Celtic salt to taste

1. Preheat oven to 350°F.
2. Trim off and wash the stems from the mushrooms. Mince the stems fine.
3. In a skillet, sauté the onions, garlic, and stems in the olive oil about 4 minutes. Add the bread crumbs and parsley and combine well. Add Celtic salt to taste.
4. Stuff the mushrooms with the mixture and brush the tops with additional olive oil.
5. Bake in oven in a shallow baking pan about 15 minutes or until nicely browned.

*Jaclyn's is recommended.

Butternut Squash Soup

1 medium butternut squash, peeled and seeded, then cut into cubes

1 Granny Smith apple, cored and chopped (skin can be left on if organic)

8 cups filtered or bottled water or stock, made by dissolving 4 vegetable boullion cubes in the water, which has been warmed

1 medium onion, thinly sliced

1 clove garlic, minced

3 tablespoons honey

1 teaspoon cinnamon

1 cup unsweetened soy milk

Celtic salt to taste

½ cup raw pumpkin seeds

1. In a large soup pot, combine squash, stock or water, onions, and garlic. Bring to a boil and then simmer, covered, for 45 minutes.
2. Puree half the soup in a blender and pour back into soup pot.
3. Using a whisk, add the soy milk, honey, cinnamon, and salt to taste. Heat on simmer for 5 to 10 minutes, stirring occasionally.
4. Serve in soup bowls with 1 tablespoon pumpkin seeds as garnish.

Walnut Flax Bars

3 tablespoons walnut oil

1 tablespoon blackstrap molasses

⅓ cup raw sugar

¾ cup chopped walnuts

2 organic eggs

2 teaspoons vanilla extract

1 cup wheat germ

⅔ cup ground flaxseeds

½ teaspoon baking powder

1. Preheat oven to 350°F.
2. Beat together the first 7 ingredients.

3. Mix together the ground flaxseeds and baking powder and fold into the beaten mixture.
4. Pour mixture into an 8-inch greased shallow square pan.
5. Bake in oven for 30 minutes or until firm to the touch.
6. Cut into squares and serve.

Barley Pilaf

1 cup whole barley grain, soaked for at least 4 hours
    in filtered or bottled water and drained

2 tablespoons extra-virgin olive oil

1 medium onion, chopped

1 teaspoon fresh or ½ teaspoon dried tarragon

3 cups filtered or bottled water or miso stock (use recipe that follows)

Celtic salt to taste

2 tablespoons minced fresh parsley as garnish

1. In a large heavy pot, sauté onion in olive oil about 5 minutes.
2. Add the barley and cook for about 2 minutes. Add the tarragon, stock or water, and Celtic salt to taste.
3. Bring to a boil, then simmer covered for about 45 minutes or until the barley is tender. Add more water and cook longer if needed.
4. Garnish each serving with parsley.

Miso Stock

¼ cup barley miso paste

4 cups filtered or bottled water

Bring 4 cups water to a boil. Allow to cool for 3 minutes. Add miso paste and stir to dissolve thoroughly.

Apple and Walnut Soufflé

⅔ cup whole wheat pastry flour

3 teaspoons baking powder

2 eggs

½ cup raw sugar

3 teaspoons vanilla

2 cups diced apples

1 cup walnuts chopped finely

¼ teaspoon Celtic salt

plain yogurt for garnish

1. Preheat oven to 350°F.
2. In a large bowl, beat eggs. Add sugar and vanilla and beat well.
3. Mix flour, baking powder, and salt and fold into egg mixture. Add apples and walnuts.
4. Transfer to deep baking dish and bake in oven for 45 minutes.
5. Serve hot with a dab of yogurt.

Tarragon Chicken with Lemon

1 chicken, cut into eighths

1 large onion, sliced

5 crushed garlic cloves

juice of 2 lemons

½ teaspoon cayenne pepper

1 tablespoon fresh or 1 teaspoon dried tarragon

1. Preheat oven to 325°F.
2. Scatter onion slices over large roasting pan.
3. Add chicken pieces.
4. Mix garlic with lemon juice and add the tarragon and pepper. Pour over chicken pieces.
5. Bake in oven for about 1½ hours.

Baked Trout

3 or 4 1-pound trout fillets

4 large lettuce or other large green leaves

2 teaspoons minced garlic

juice of 1 lemon

¼ cup fresh chopped parsley leaves

¼ cup extra-virgin olive oil

Celtic salt to taste

1. Preheat oven to 350°F.
2. Line a large baking pan with green leaves and drizzle with 1 tablespoon olive oil.
3. Distribute half of the garlic and half of the parsley over leaves.
4. Top with fish fillets. Drizzle the remaining olive oil over fish and top with remaining garlic and parsley. Squeeze lemon juice over fillets.
5. Bake in oven for about 20 minutes or until the edges of the fillets flake with a fork. Do not overcook.

Poached Pears with Walnuts

4 pears of choice (ripe but not too soft)

1 cup filtered or bottled water

1 cup mirin*

1 cinnamon stick

2 tablespoons maple syrup

1 lemon, sliced

¼ cup whole raw walnuts

1. Preheat oven to 325°F.
2. Remove the core from the nonstem end of the pears; leave the stems on. Place in a baking dish.
3. In an uncovered saucepan, bring water, mirin, maple syrup, lemon slices, and cinnamon stick to a boil. Immediately reduce heat to low and simmer for 10 minutes. Pour over pears and cover dish.

4. Bake pears in oven for about 20 minutes or until pears are tender.
5. Serve pears with a ladle of syrup and surround each pear with fresh raw walnuts.

Green Salad with Poppy Seed Dressing

salad greens
¼ cup extra-virgin olive or walnut oil
3 teaspoons poppy seeds
2 cloves of garlic, minced
1 tablespoon each brown rice vinegar and umeboshi vinegar*
1 teaspoon Dijon mustard

1. Combine all ingredients except greens into a jar and shake to combine well.
2. Serve over greens.

*Available at health food stores.

Tomato Basil Omelet

4 organic eggs
1 ripe tomato, chopped
½ onion, chopped
2 teaspoons fresh basil
1 teaspoon butter
1 tablespoon extra-virgin olive oil

1. Beat eggs with a fork until blended.
2. Heat olive oil and butter in a skillet. Sauté the onion for about 2 minutes and then add the tomato, sautéing for another 2 minutes. Remove vegetables from pan and reserve.

3. Add eggs to the skillet and allow to cook for 1 minute. Using a spatula, push the edges of the eggs toward the center and tilt the pan to allow the liquid portion of the egg to run to the edges of the pan. Repeat until egg is almost set, about 3 minutes total.
4. Place vegetables across one side of the omelet. Top with fresh basil. Using a spatula, fold other side over the vegetables and serve.

Split Pea Dill Soup with Pumpkin Seeds

2 cups split peas, rinsed

8 cups filtered or bottled water

3 cloves garlic, minced

2 onions, chopped

2 stalks celery, sliced

2 carrots, sliced

2–4 red potatoes, diced

1 teaspoon dill, dried, or, if available, 1 tablespoon fresh

Celtic salt to taste

1 tablespoon per person raw pumpkin seeds

1. Combine all of the ingredients except the pumpkin seeds in a large soup pot.
2. Bring to the boil, then reduce heat to simmer, lightly covered.
3. Cook until peas are tender, about 1 hour. Add more water, as needed, if soup is too thick. Soup can be pureed in blender or food processor, as desired.
4. Serve sprinkled with pumpkin seeds.

Cranberry Flaxseed Muffins

1 cup buckwheat flour

1 cup oat flour

1 cup spelt flour

2½ teaspoons baking powder

¼ cup honey

1 cup dairy or soy milk

¼ cup melted butter

1 egg, beaten

¾ cup unsweetened cranberries, fresh or frozen

½ cup flaxseeds

1. Preheat oven to 350°F.
2. Grease muffin tins or use paper liners.
3. In a large bowl, stir together the honey, milk, butter, and egg. Combine the dry ingredients together in a separate bowl and then add to wet ingredients, mixing until well combined.
4. Fold in cranberries and flaxseeds.
5. Bake in oven for 20 minutes or until a toothpick comes out clean.

Honey-Glazed Baked Wild Alaska Salmon

2 pounds fresh wild salmon

Juice of a lemon

1 tablespoon extra-virgin olive oil

1 tablespoon tamari soy sauce

3 tablespoons honey

1 tablespoon mirin*

1 tablespoon freshly grated ginger

1. Preheat oven to 375°F.
2. Place fillets in a glass baking dish.
3. Combine remaining ingredients and pour over fish. Marinate in refrigerator overnight or for at least 1 hour. Remove from refrigerator and let stand for about 15 minutes.
4. Bake in oven for about 20 minutes or until salmon is pink and tender but not overcooked.

*Available at health food stores.

Avocado Salsa

2 ripe avocados
1 tablespoon lemon juice
2 medium tomatoes, chopped
1 red onion, chopped fine
1 green pepper, chopped fine
1 clove garlic, minced
1 tablespoon extra-virgin olive oil
Celtic salt to taste

1. In a small bowl, mash meat from avocados and combine with lemon juice.
2. Add the remaining ingredients and mix to combine. A smoother consistency can be obtained by blending the contents in a blender for a few minutes.
3. Serve with assorted vegetables.

Flank Steak with Tamari Soy Sauce and Garlic

1 flank steak
¼ cup tamari soy sauce
¼ cup honey
3 garlic cloves, crushed

1. Marinate flank steak in the other ingredients for 1 hour in a broiling pan.
2. Broil on medium heat for about 3 to 4 minutes on each side or to desired doneness.

Quinoa and Beef Stuffed Red Peppers

1 cup quinoa, cooked according to package directions

1 medium onion, chopped

½ pound lean ground beef

1 cup chopped tomatoes, fresh if possible

1 tablespoon extra-virgin olive oil

fresh basil to taste

4–5 red peppers, cored and seeds removed

1. Preheat oven to 375°F.
2. Sauté beef and onions in olive oil in skillet. Drain excess fat.
3. Combine beef mixture, tomatoes, and quinoa, cutting in a bit of fresh basil.
4. Place peppers in baking dish and fill three-quarters full. Pour 1 inch of filtered or bottled hot water around peppers and cover.
5. Bake in oven for 1 hour or until peppers are soft.

Broiled Venison (or Beef) Cubes

1 pound venison or beef cubes

½ cup organic wine

3 garlic cloves, chopped

1 tablespoon miso paste (barley)

2 tablespoons extra-virgin olive oil

1. Marinate cubes overnight in the other ingredients.
2. Drain meat, saving marinade mixture to baste cubes.
3. Thread meat on kebab sticks and broil for 5 to 10 minutes, basting with marinade.

Sprouted-Grain French Toast

2 eggs, beaten
2–3 slices Ezekiel sprouted-grain bread
Celtic salt to taste
organic butter
all-fruit unsweetened jam

1. Heat butter in large skillet.
2. Dip each piece of bread into beaten eggs and immediately cook in skillet until browned on both sides.
3. Serve immediately with jam.

Black Bean Soup with Pumpkin Seed Garnish

1 cup black beans, soaked overnight, or 1 can low-sodium black beans*
7 cups filtered or bottled water
1 tablespoon extra-virgin olive oil
2 cloves garlic, minced
1 large onion, minced
2 stalks of celery, sliced thinly
2 carrots, sliced thinly
¼ teaspoon ground cumin
Celtic salt to taste
¼ cup raw pumpkin seeds

1. Drain soaked beans or canned beans.
2. In a large soup pot, add soaked beans and water and simmer after bringing them to a boil, partially covered, for about 2–3 hours. Skip this step if you are using the canned beans.
3. Sauté onion, garlic, carrot, and celery in olive oil until tender in skillet. Add beans, cumin, and water and simmer for another 30 minutes.
4. Puree soup in blender or food processor.
5. Serve soup in bowls garnished with pumpkin seeds.

*Eden black beans are low in sodium and are found in health food stores.

Fruit and Walnut Chews

¾ cup walnuts
⅔ cup raisins
¾ cup dates
⅔ cup dried unsweetened coconut

1. In a food processor, coarsely mince raisins, dates, and walnuts.
2. Add coconut. Work mixture with clean fingers until it binds to-gether.
3. Form into rolls and cut into bite-sized pieces.
4. Serve cold or at room temperature.

Broccoli Frittata

5 eggs
1 tablespoon filtered or bottled water
1 small onion, chopped
3 tablespoons olive oil
1 cup broccoli florets, cut into small pieces
3 shiitake mushrooms, sliced
½ cup grated raw milk cheese such as Monterey Jack
3–5 sprigs of parsley

1. Preheat oven to 350°F.
2. In a large ovenproof skillet, at low temperature, sauté onion and mushrooms until soft in 1 tablespoon olive oil. Add broccoli and continue to sauté for a few minutes longer.
3. Beat eggs with water. Add vegetables and cheese.
4. Add and heat remaining 2 tablespoons olive oil in skillet. Add egg mixture and cook on medium-low heat about 10 minutes or until bottom is firm.
5. Transfer frittata to oven and bake for about 10 to 20 minutes or until top is no longer runny. Check often to prevent overcooking.
6. Serve garnished with parsley.

Chicken and Ginger Sauté over Spelt Pasta

1 pound chicken breasts, cut into 1-inch cubes

2 tablespoons extra-virgin olive oil

2 garlic cloves, minced

3 tablespoons fresh ginger root, peeled and minced or grated

¼ cup dry white wine or mirin

¼ cup filtered or bottled water

2 tablespoons tamari soy sauce

Spelt pasta, cooked

1. In a large skillet, sauté garlic and ginger in olive oil for 2 minutes on low heat.
2. Add wine or mirin and water and cook until liquid has been reduced by half. Add the soy sauce and cook for 2 minutes, stirring.
3. Add the chicken and sauté on medium, stirring often, until cooked through, about 6 minutes.
4. Serve over spelt pasta.

Nut and Seed Butters

Most organic nut and seed butters can be purchased in a health food store. To make your own, use:

1 cup raw organic nuts or seeds

Filtered or bottled water

Grind nuts or seeds in a coffee grinder or blender until a coarse or fine texture is achieved, as desired. Add just enough water to make a paste. Store in refrigerator.

Walnut Quick Bread

Crust

½ cup soft butter

1 cup whole wheat pastry flour

1 tablespoon honey

1 teaspoon ground cinnamon

Filling

3 eggs, beaten

¾ cup walnuts, chopped fine

¼ cup whole wheat pastry flour

1 teaspoon baking powder

1 cup shredded dried unsweetened coconut

¼ cup honey

1. Preheat oven to 350°F.
2. Mix together the crust ingredients and press into a square 10-inch pan.
3. Mix filling ingredients and spread on top of crust.
4. Bake in oven for 30 minutes or until lightly browned on top.
5. Cut into slices when cooled and serve.

Salmon Cakes

1 large can salmon

¼ cup whole wheat bread crumbs

juice of 1 lemon

2 tablespoons onion, finely minced

1 teaspoon Dijon mustard

1 egg, beaten

Celtic salt to taste

3 tablespoons extra-virgin olive oil

1. Flake fish in a bowl with bread crumbs.
2. Add remaining ingredients except olive oil, mixing until well combined.
3. Form into cakes the size of hamburgers.
4. Heat olive oil in a large skillet. Add cakes and cook over medium heat about 5 minutes. Flip over and cook until nicely browned.

Chicken Stir-fry with Walnuts and Vegetables

1 pound deboned chicken, cut into strips
2 tablespoons extra-virgin olive oil
1 red pepper, chopped
1 onion, chopped
2 cups broccoli florets, chopped
1 cup snowpeas
tamari soy sauce to taste
fresh bean sprouts

1. In a large skillet, sauté onion, pepper, broccoli, and snow peas in olive oil until crisp-tender.
2. Add chicken and sauté, stirring until chicken is cooked through.
3. Serve with bean sprouts and tamari soy sauce.

Broiled Tofu with Walnut Sauce

1 pound firm tofu, cut into 6½-inch-thick pieces
1 tablespoon extra-virgin olive oil
½ cup walnuts
2 tablespoons barley miso dissolved in ¼ cup filtered or bottled warmed water
1 teaspoon honey

1. In a baking pan, arrange tofu slices in a single layer and brush with olive oil.

2. To make sauce, grind walnuts in a coffee grinder. In a bowl, combine ground walnuts with miso and water and honey. Mix well to form a thick sauce.
3. Broil tofu about 4 minutes on each side.
4. Spread with sauce and broil for an additional minute.

Tilapia Fish Chowder

1 pound tilapia fillets, cut into 1-inch pieces
1 tablespoon olive oil
2 medium red potatoes, chopped
1 clove garlic, minced
3 carrots, sliced
2 large onions, sliced
1 tablespoon fresh dill or 1 teaspoon dried
1 tablespoon whole wheat flour
2 cups filtered or bottled water, brought to a boil
½ cup dry white wine
1 cup soy or dairy milk (optional)
parsley for garnish

1. In a large, heavy pot, sauté onions, potatoes, garlic, and carrots in olive oil for 5 minutes. Add the flour and cook for a few additional minutes, stirring with a wooden spoon. Then add dill.
2. Add boiling water and simmer, covered, for 15 minutes.
3. Add the fish cubes and the wine. Simmer, covered, for about 10 minutes.
4. Serve as is or heat with milk. Garnish with parsley.

Spinach Salad with Lemon Miso Dressing

1 pound washed and torn spinach or baby spinach
½ cup extra-virgin olive oil
juice of 1 lemon
1 tablespoon sweet white miso
1 tablespoon tamari soy sauce
Celtic salt to taste

1. Place spinach in a large serving bowl.
2. Add enough dressing to moisten leaves and serve.

Lamb Curry with Walnuts

3 pounds lamb shoulder, cut into 2-inch cubes
4 onions, sliced
2 cloves garlic, minced
3 tablespoons extra-virgin olive oil
3 tablespoons curry powder
3 cups filtered or bottled water
2 apples, cored, and chopped (if organic, need not be peeled)
1 cup walnuts, chopped
2 lemons, sliced
4 tablespoons raisins
½ cup unsweetened shredded coconut
1 tablespoon brown rice syrup

1. In a large, heavy pot, sauté onions and garlic in olive oil until onions are tender. Add lamb cubes and sauté 10 minutes, stirring frequently.
2. Add curry powder and simmer 5 minutes. Add remaining ingredients and pour water over all. Bring mixture to a boil, cover pot, and reduce heat to simmer. Simmer 1 hour or until lamb is cooked through.

Baked Eggs Françoise

8 ripe tomatoes, hollowed out

8 omega 3–enriched eggs

4 tablespoons fresh parsley, snipped

2 teaspoons garlic powder

Celtic salt to taste

3 tablespoons extra-virgin olive oil

3 tablespoons fresh basil, snipped

¼ cup grated raw milk cheese such as Gruyère

1. Preheat oven to 400°F.
2. Grease a large baking dish with 1 tablespoon olive oil and place tomatoes in it. Sprinkle with garlic powder, parsley, and Celtic salt. Bake in oven for 10 minutes.
3. Remove tomatoes from oven. Break one egg into each tomato; sprinkle with olive oil and basil. Bake for another 20 minutes. Sprinkle with cheese and bake for about 10 minutes or until cheese is melted.

Pumpkin Soup Garnished with Seeds

2 pounds fresh pumpkin, peeled and cut into 1-inch cubes

2 tablespoons sweet white miso dissolved in 6 cups filtered or bottled water

2 large onions, sliced

2 teaspoons grated fresh ginger root

2 teaspoons kuzu* dissolved in ¼ cup filtered or bottled water

1 cup heavy cream, whipped

pumpkin seeds

1. In a large soup pot heat the miso broth. Add pumpkin, ginger, and onions. Cover and simmer soup for about 30 minutes.
2. Puree the soup in a blender and return to pot. Add kuzu and water and cook, stirring, to thicken the soup.

3. Serve sprinkled with a dollop of whipped cream topped with pumpkin seeds.

*Kuzu, a starch from the kuzu plant used in Japanese cuisine as a thickener, can be found in health food stores.

Endive and Radicchio Salad with Olive and Walnut Oils

1 medium head radicchio, leaves removed and cut into bite-sized pieces

2 endives, leaves removed

2 tablespoons extra-virgin olive oil

2 tablespoons walnut oil

juice of 1 lemon

1 garlic clove, crushed

1. Make dressing by mixing oils, lemon juice, and garlic clove in a bowl or jar.
2. Decorate each serving plate with radicchio leaves. Scatter endive over radicchio.
3. Serve with dressing.

Apple and Walnut Squares

2 Granny Smith apples, diced (if organic, no need to remove skin)

1 cup walnuts, chopped

1 cup pitted dates, chopped

1 cup whole wheat pastry flour

1½ teaspoons baking powder

¼ cup honey

1 tablespoon melted butter

1 organic egg, beaten

1. Preheat oven to 400°F.
2. In a large bowl, mix all ingredients until well combined.

3. Spread mixture into an 8-inch square greased pan. Bake in oven for 30 minutes or until golden.
4. Cool, then cut into squares.

Leaf and Frisee Lettuces with Lemon Walnut Vinaigrette

Use bagged mesclun that contains both leaf and frisee lettuces.

Dressing
2 tablespoons extra-virgin olive oil
2 tablespoons walnut oil
2 tablespoons balsamic vinegar
1 garlic clove, crushed

1. Combine dressing ingredients in a jar and shake well.
2. Pour over mesclun to serve.

Millet Pancakes

½ cup cooked millet
½ cup cooked brown rice
2 tablespoons grated carrot
2 tablespoons grated onion
cornmeal for coating
sesame oil for sautéing
applesauce

1. Combine the rice, millet, grated carrot, and onion. Form cakes, using clean wet hands.
2. Coat each side of the cakes in cornmeal. (Use a plate or shallow bowl to coat the cakes.)

3. Heat 3 tablespoons of oil in a skillet and sauté the pancakes on both sides until golden. Drain on paper towels.
4. Serve with applesauce if desired.

Chef's Salad with Chicken and Pumpkin Seed Vinaigrette

salad greens
2 cups shredded cooked chicken

Dressing
4 tablespoons pumpkin seed oil
2 tablespoons balsamic vinegar
1 garlic clove, crushed

Dress greens with chicken and top with pumpkin seed vinaigrette.

Walnut Pâté

1½ cups cooked green beans, blended in blender
2 hard-cooked eggs, grated
½ cup walnut, ground
3 tablespoons extra-virgin olive oil
1 tablespoon mirin or white wine
Celtic salt to taste

1. In a skillet, sauté onion in 1 tablespoon olive oil.
2. Combine remaining ingredients, adding the sautéed onion.
3. Chill. Serve with crackers or cut-up vegetables.

Chapter 4

For information on drug-induced nutrient depletion, see *Drug-Induced Nutrient Depletion Handbook,* by Ross Pelton, James B. LaValle, Ernest B. Hawkins, and Daniel Krinsky, published by Lexi-Comp, Inc., Hudson Ohio, and Natural Health Resources, Cincinnati, Ohio.

The book can be purchased on amazon.com.

Chapter 5

To find a directory of merchants who sell *organic products* on the Internet or via mail order, go to:
http://directory.google.com/Top/Shopping/Food/Organic.

For information on *gluten intolerance* and links to websites and articles, go to: www.gluten-free.org or gfcfdiet.com.

Chapter 6

Essential fatty acids: At the Perlmutter Center, we use essential fatty acids from Nordic Naturals. Their oils are manufactured in a nitrogen environment, which prevents damaging oxidation. Their products undergo third-party testing, which demonstrates positively no PCBs or heavy metals, exceeding FDA purity levels by 400 times. Their omega oils are an excellent source of DHA and come in a variety of pleasant, nonfishy natural fruit flavors in both liquid and capsules. Nordic Natural Products can be purchased from www.nordicnaturals.com, or by phone 800-662-2544.

Neuromins® brand DHA supplements contain *Martek DHA™*, an all-natural vegetarian source of DHA derived from algae. This is an excellent source of DHA, free of environmental toxins. Neuromins® is available at many health food stores. For more information, call the DHA Information Center at 1-888-OK-BRAIN (888-652-7246) or visit www.DHADepot.com.

All of the products mentioned in the Better Brain Book can be purchased separately at health food stores, pharmacies, and even many discount stores. Although I can't vouch for every product, I can recommend some well-known brands, many of which are available from your local health food stores, including Ecological Formulas/Cardiovascular Research, Nature Made, and the GNC brand. Q-Gel makes an excellent Co-Q10.

Brain Sustain and Brain Sustain Neuroactives are advanced comprehensive nutritional products specifically designed by David Perlmutter, M.D., for maintenance and performance enhancement. Each scoop (20 grams) of Brain Sustain provides the following: vitamin C (as calcium ascorbate), 200 milligrams; vitamin D, 200 IU; vitamin E (as d-alpha tocopheryl), 200 IU; niacin, 50 milligrams; vitamin B6 (as pyridoxal 5'-phosphate), 50 milligrams; folate (folic acid), 400 micrograms; vitamin B12 (as methylcobalamin), 100 micrograms; N-acetyl-cysteine, 200 milligrams; phosphatidylserine, 50 milligrams; acetyl-L-carnitine, 200 milligrams; lipoic acid, 60 milligrams; coenzyme Q10, 30 milligrams; ginkgo biloba extract (leaf), 24 percent ginkgo heterosides, 30 milligrams. This product contains no wheat, soy, or corn. *Two capsules of Brain Sustain Neuroactives provide the following:* N-acetyl-cysteine, 200 milligrams; phosphatidylserine, 50 milligrams; acetyl-L-carnitine, 200 milligrams; lipoic acid, 40 milligrams; coenzyme Q10, 30 milligrams; ginkgo biloba,

30 milligrams. For more information, go to www.nutritionals.com or call 1-800-530-1982.

Chapter 7

For information on yoga, go to:

> www.yoga.com, or call 866-266-9643. Everything about yoga.
> www.yogasite.com/: this site can help you locate a yoga teacher and offers other pertinent information.
> www.yogajournal.com/: excellent consumer magazine about yoga.
> www.yogadirectory.com/: this site can help you locate a yoga teacher in your area and offers other pertinent information.
> www.yogafinder.com

Chapter 8

NATURAL BUG REPELLENT WEBSITES AND OTHER RESOURCES

www.bitestop.com

Bite Stop products are a natural alternative to DEET-based repellents.

www.kokogm.com

Official website of Kokopelli's Green Market. They sell a variety of organic products, including natural insect repellents for humans as well as pets. Their natural insect repellents are a blend of plant oils, plant extracts, and plant derivatives.

www.herbalremedies.com

Offers mosquito and insect repellent information and products as well as biting insect information.

www.planetnatural.com/barriersrepellents

Natural insect repellent products, including natural squirrel, raccoon, and deer repellents

www.pestrepellents.com

Offers user-friendly, environment-friendly, ozone-friendly natural insect repellents, including the revolutionary "Flies Be Gone Fly Trap" that contains no toxins, poisons, or insecticides.

www.naturalbabyproducts.com

Buzz Away sells an all-natural insect repellent. Independent lab tests prove that Buzz Away keeps the bugs at bay for hours, and because Buzz Away is 100 percent DEET-free and natural, it is safe to use on children.

www.equuskreen.com

Skreen Products—available online. All-natural insect repellent and soothes itching and discomfort from existing bites and stings.

www.thebackpacker.com

All-terrain, all-natural insect repellents.

www.wisementrading.com

Natural insect repellents, including Neem Aura Naturals: Herbal Outdoor Spray. Neem is one of those select herbs that have been used for centuries for their multipurpose properties. It is known in India as the "village pharmacy." Neem Aura Herbal Outdoor Spray allows us to enjoy the outdoors without toxic chemicals or insect repellents.

www.naturalfoodmerchandiser.com

Information on natural insect repellents.

www.heavenscentaroma.com

Make your own natural insect repellents using essential oils.

www.quantumhealth.com

Sells Buzz Away natural insect repellent sprays and other products. Also has articles online about natural insect control.

www.smartshield.com/sunscreen

Sells SPF30 sunscreen spray with natural insect repellent. Waterproof and sweatproof.

Books

> *Natural Insect Repellents for Pets, People, and Plants* by Janette Grainger and Connie Moore (available at www.nelsonbooks.com and amazon.com)

Handbook of Natural Pesticides: Insects Attract Repellents, vol. 6, by D. Morgan (available through amazon.com)

WEBSITES ON PESTICIDE-FREE GARDENING AND ORGANIC GARDENING

www.eap.mcgill.ca/publications
Articles on ecological lawn management.

www.organicgardening.com
Information on safe pest control.

www.pesticide.org
Home page for the Northwest Coalition for Alternatives to Pesticides.

www.gardensalive.com
Environmentally responsible products for gardening that do not contain pesticides.

www.pesticidefreeyards.org
Sponsored by the Sierra Club.

BOOKS ON ORGANIC GARDENING:

Rodales's Illustrated Encyclopedia of Organic Gardening, edited by Pauline Pears (Rodale Press, 2002; available through www.amazon.com)
How to Get Your Lawn and Garden Off Drugs, by Carole Rubin (Harbour Publishing, 2002; available through www.amazon.com)

For information on testing *lead levels* in your drinking water, go to:

www.leadingtesting.org
www.osha.gov/SLTC/leadtest/intro.html

For information on testing *mercury levels* in your drinking water:
www.cleanwateraction.org/mercury/facts.html

For information on *mercury-free dentistry,* contact: the International Academy on Oral Medicine and Toxicology, Box 608010, Orlando, FL 32860-5831; 863-420-6373.

For information on where to locate a doctor who performs *chelation therapy,* contact: American Board of Chelation Therapy, 70 West Huron Street, Chicago, IL 60610; 312-266-7246.

For information on *reverse osmosis water filtration* systems, go to: www.theolivebranch.com/water/ro.htm.

For information on the safe use of *cell phones,* check out this U.S. government–sponsored website: www.fda.gov/cellphones.

For information on common *household products* that contain aluminum or other toxins, check out the Household Products Database at: http://householdproducts.nlm.nih.gov/products.htm.

Chapter 9

BrainBuilder, a computer-based software program designed to enhance mental performance, can be purchased online from www.Advanced Brain.com, or by phone 800-530-1982. It is also available from www.iNutritionals.com, or by phone 800-530-1982.

ThinkFAST-Neurobic Software is another excellent computer-based mental enhancement program. It can be purchased online from www.brain.com.

Chapter 10

Home antioxidant test kits mentioned in chapter 10 are available from the following companies:

Antioxidant Check by Body Balance: www.ULPTEST.com, or call 800-530-1982.

Vespro Free Radical Test Kit: www.Vespro.com or call 1-800-438-4894.

OxyStress from North American Pharmacal: www.4yourtype.com, or call 877-ABO TYPE.

The Glutathione Protocol: Information for You and Your Doctor

OUR PROTOCOL FOR intravenous glutathione is fairly simple, but it should be administered, at least initially, by a qualified health professional. At the Perlmutter Center, we usually teach a spouse or family member to provide glutathione injections at home.

Glutathione is easily obtained from several pharmacies in the United States. (See the list that follows.) We use liquid glutathione, not reconstituted powder. Typically, a month's worth of glutathione is mailed for overnight delivery in a light-proof foil container on dry ice. Ampules usually contain 200 milligrams of glutathione per cubic centimeter; a 7 cubic centimeter ampule provides 1,400 milligrams of glutathione. We mix the glutathione with 5 cc of sterile saline solution, and the mixture is injected over a ten-minute period. Most patients require injections three times a week. We have used doses up to 2,000 milligrams every other day in some patients to achieve the best results. We continue these injections indefinitely. In fact, most of our patients are reluctant to reduce or discontinue glutathione therapy because they feel so much better after their treatments.

Glutathione should be administered as follows.

1. The glutathione is mixed with 5 cc of sterile saline solution.
2. The solution is then injected through a 23- or 25-gauge butterfly catheter intravenously over a 10-minute period.
3. Many patients, especially those with difficult veins, choose to have intravenous access ports inserted. This allows for repeated glutathione treatments without repeated needle sticks.

To find a doctor who performs IV glutathione therapy, contact the American Academy for the Advancement in Medicine at www.ACAM. org, or call 949-583-7666.

For a complete instructional video on glutathione administration, contact iNutritionals at 1-800-530-1982, or wwwiNutritionals.com. The instructional video is designed to be used in conjunction with your physician.

Injectable glutathione is available from Wellness Health and Pharmacy, 3401 Independence Drive, Suite 231, Birmingham, AL 35209; 1-800-227-2627, and Abrams-Royal Pharmacy, 8220 Abrams Road, Dallas, TX 75231; 1-214-349-8000.

Oxidative Stress: Role of Free Radicals in Brain Degeneration and Functional Decline

Berr, C. Oxidative stress and cognitive impairment in the elderly. *J Nutr Health Aging* 6(4): 261–6, 2002.

Cotman, C.W., et al. Brain aging in the canine: A diet enriched in antioxidants reduces cognitive dysfunction. *Neurobiol Aging* 23(5): 809, 2002.

Fukui, K., et al. Cognitive impairment of rats caused by oxidative stress and aging, and its prevention by vitamin E. *Ann N Y Acad Sci* 959: 275–84, 2002.

Martin, A., et al. Effects of fruits and vegetables on levels of vitamins E and C in the brain and their association with cognitive performance. *J Nutr Health Aging* 6(6): 392–404, 2002.

Martin, A., et al. Roles of vitamins E and C on neurodegenerative diseases and cognitive performance. *Nutr Rev* 60(10), pt 1: 308–26, 2002.

Morris, M.C., et al. Dietary intake of antioxidant nutrients and the risk of incident Alzheimer disease in a biracial community study. *JAMA* 287(24): 3230–7, 2002.

Morris, M.C., et al. Vitamin E and cognitive decline in older persons. *Arch Neurol* 59(7): 1125–32, 2002.

Rao, A.V., Balachandran, B. Role of oxidative stress and antioxidants in neurodegenerative diseases. *Nutr Neurosci* 5(5): 291–309, 2002.

Reis, E.A., et al. Pretreatment with vitamins E and C prevents the impairment of memory caused by homocysteine administration in rats. *Metab Brain Dis* 17(3): 211–7, 2002.

Zhang, S.M., et al. Intakes of vitamins E and C, carotenoids, vitamin supplements, and PD risk. *Neurology* 59(8): 1161–9, 2002.

Alzheimer's Disease Risk Factors

Ahlbom, A., et al. Neurodegenerative diseases, suicide, and depressive symptoms in relation to EMF. *Biolelectromagnetics* 22: S132–43, 2001.

Anonymous. The Canadian Study of Health and Aging: Risk factors for Alzheimer's disease in Canada. *Neurology* 44: 2073–80, 1994.

Barrett-Connor, E., et al. Weight loss precedes dementia in community-dwelling older adults. *J Am Geriatr Soc* 44: 1147–52, 1996.

Beardsley, T. Say that again. *Scientific American* 277: 20, 1997.

Breteler, M.M., et al. Epidemiology of Alzheimer's disease. *Epidemiol Rev* 14: 59–82, 1992.

Broe, G.A., et al. A case control study of Alzheimer's disease in Australia. *Neurology* 40: 1698–1707, 1990.

Fabrigoule, C., et al. Social and leisure activities and risk of dementia: A prospective longitudinal study. *J Am Geriatr Soc* 43: 485–90, 1995.

Flynn, R.R., et al. Organic solvents and presenile dementia: A case referent study using death certificates. *Br J Ind Med* 44: 259–62, 1987.

Francese, T., et al. Effects of regular exercise on muscle strength and functional abilities of late stage Alzheimer's residents. *American Journal of Alzheimer's Disease* 12: 122–7, 1997.

Friedland, R.P., et al. Patients with Alzheimer's disease have reduced activities in midlife compared with healthy control-group members. *Proc Natl Acad Sci USA* 98: 3440–5, 2001.

Graves, A.B., et al. Cognitive decline and Japanese culture in a cohort of older Japanese Americans in King County, Washington. *J Gerontol B Psychol Sci Soc Sci* 54: S154–61, 1999.

Graves, A.B., et al. Occupational exposure to electromagnetic fields and Alzheimer's disease. *Alzheimer Dis Assoc Disord* 13: 165–70, 1999.

Henderson, A.S., et al. Environmental risk factors for Alzheimer's disease: Their relationship to age of onset and to familial or sporadic types. *Psychol Med* 22: 429–36, 1992.

Jost, B.C., et al. The evolution of psychiatric symptoms in Alzheimer's disease: A natural history study. *J Am Geriatr Soc* 44: 1078–81, 1996.

Kolanowski, A.M., Restlessness in the elderly: The effect of artificial lighting. *Nursing Research* 39: 181–3, 1990.

Kondo, K., et al. A case-control study of Alzheimer's disease in Japan: A significance of lifestyles. *Dementia* 5: 314–26, 1994.

Kudoh, A., et al. Response to surgical stress in elderly patients and Alzheimer's disease. *Can J Anaesth* 46: 247–52, 1999.

Kukull, W.A., et al. Solvent exposure as a risk factor for Alzheimer's disease: A case control study. *Am J Epidemiol* 141: 1059–71, 1995.

Laurin, D., et al. Physical activity and risk of cognitive impairment and dementia in elderly persons. *Arch Neurol* 58: 498–524, 2001.

Leverenz, J.B., et al. Effect of chronic high dose cortisol on hippocampal neuronal number in aged nonhuman primates. *J Neuroscience* 19: 2356–61, 1999.

Mayeaux, R., et al. Reduced risk of Alzheimer's disease among individuals with low caloric intake. *Neurology* 52: A296, 1999.

Murialdo, G., et al. Relationships between cortisol, dehydroepiandrosterone sulphate and insulin-like growth factor 1 system in dementia. *J Endocrinol Invest* 24: 139–46, 2001.

Newcomer, J.W., et al. Decreased memory performance in healthy humans induced by stress-level cortisol treatment. *Arch Gen Psychiatry* 56: 527–33, 1999.

Okamoto, K., et al. Sociomedical and life style risk factors of senile dementia, determined in a nested case-control study. *Nippon Ronen Igakkai Zaashi* 31: 604–9, 1994.

Parshad, R., et al. Effect of DNA repair inhibitors on the in vitro test for Alzheimer's disease. *J Am Geriatr Soc* 46: 1331–2, 1998.

Raiha, I., et al. Environmental differences in twin pairs discordant for Alzheimer's disease. *J Neurol Neurosurg Psychiatry* 65: 785–7, 1998.

Rasmusson, S., et al. Increased glucocorticoid production and altered cortisol metabolism in women with mild to moderate Alzheimer's disease. *Biol Psychiatry* 49: 547–2, 2001.

Reding, M., et al. Depression in patients referred to a dementia clinic: A three-year prospective study. *Arch Neurol* 42: 894–6, 1985.

Shen, Y., et. al. A case control study of risk factors on Alzheimer's disease: Multicenter collaborative study in China. *Chung Hua Shen Ching Ching* 25: 284–7, 1992.

Shimamura, K., et al. Environmental factors possibly associated with onset of senile dementia. *Nippon Koshu Eisei Zaashi* 45: 203–12, 1998.

Shulte, P.A., et al. Neurodegenerative diseases: Occupational occurrence and potential risk factors, 1982 through 1991. *Am J Public Health* 86: 1281–8, 1996.

Sobel, E., et al. Elevated risk of Alzheimer's disease among workers with likely electromagnetic field exposure. *Neurology* 47: 1477–1, 1996.

Sobel, E., et al. Occupations with exposure to electromagnetic fields: A possible risk factor for Alzheimer's disease. *Am J Epidemiol* 142: 515–24, 1995.

Speck, C.E., et al. History of depression as a risk factor for Alzheimer's disease. *Epidemiology* 6: 366–9, 1995.

Tsolaki, M., et al. Risk factors for clinically diagnosed Alzheimer's disease: A case-control study of a Greek population. *Int Psychogeriatr* 9: 327–41, 1997.

Umegaki, H., et al. Plasma cortisol levels in elderly female subjects with Alzheimer's disease: A cross sectional and longitudinal study. *Brain Res* 881: 241–3, 2000.

Weiner, M.F., et al. Cortisol secretion and Alzheimer's disease progression. *Biol Psychiatry* 42: 1030–8, 1997.

Weiner, M.F., et al. Cortisol secretion and Alzheimer's disease progression: A preliminary report. *Biol Psychiatry* 34: 158–61, 1993.

Young, D., et al. Environmental enrichment inhibits spontaneous apoptosis, prevents seizures, and is neuroprotective. *Nature Medicine* 5: 448–53, 1999.

Homocysteine

Boers, G.H.J., Smals, A.G.H., Trijbels, F.J.M., et al. Heterozygosity for homocystinuria in premature peripheral and cerebral occlusive arterial disease. *N Engl J Med* 313: 709–15, 1985.

Budge, M., et al. Plasma total homocysteine and cognitive performance in a volunteer elderly population. *Ann NY Acad Sci* 407–10, May 20, 2000.

Kang, S.S., Wong, P.W., Malinow, M.R. Hyperhomocysteinemia as a risk factor for occlusive vascular disease. *Ann Rev Nutr* 12: 279–98, 1992.

McCully, K.S. Vascular pathology of homocysteinemia: Implications for the pathogenesis of arteriosclerosis. *Am J Pathol* 56: 111–28, 1969.

Miller, J.W., et al. Effect of L-dopa on plasma homocysteine in PD patients: Relationship to B-vitamin status. *Neurology* 60: 1125–9, 2003.

Muller, T., Werne, B., Fowler, B. Nigral endothelial dysfunction, homocysteine, and Parkinson's disease. *Lancet* 354: 126–7, July 10, 1999.

Polyak, Z., et al. Hyperhomocysteinemia and vitamin score: correlations with silent brain ischemic lesions and brain atrophy. *Dement Geriatr Cogn Disord* 16(1): 39–45, 2003.

Prins, N.D., et al. Homocysteine and cognitive function in the elderly: The Rotterdam Scan Study. *Neurology* 59(9): 1375–80, November 12, 2002.

Ravaglia, G., et al. Homocysteine and cognitive function in healthy elderly community dwellers in Italy. *Am J Clin Nutr* 77(3): 668–73, March 2003.

Rogers, J.D., et al. Elevated plasma homocysteine levels in patients treated with levodopa: Association with vascular disease. *Arch Neurol* 60(1): 59–64, January 2003.

Sachdov, P.S., et al. Relationship between plasma homocysteine levels and brain atrophy in healthy elderly individuals. *Neurology* 58: 1539–41, May 28, 2002.

Selhub, J., et al. Association between plasma homocysteine concentrations and extracranial carotid-artery stenosis. *N Engl J Med* 332: 286–91, 1995.

Selhub, J., et al. Vitamin status and intake as primary determinants of homocysteinemia in an elderly population. *JAMA* 270: 2693–8, 1993.

Selley, M.L., et al. Increased concentrations of homocysteine and asymmetric dimethylarginine and decreased concentrations of nitric oxide in the plasma of patients with Alzheimer's disease. *Neurobiol Aging* 24(7): 903–7, November 2003.

Seshadri, S., et al. Plasma homocysteine as a risk factor for dementia and Alzheimer's disease. *N Engl J Med* 346: 476–83, February 14, 2002.

Teunissen, C.E., et al. Homocysteine: A marker for cognitive performance? A longitudinal follow-up study. *J Nutr Health Aging* 7(3): 153–9, 2003.

Ubbink, J.B., et al. Vitamin requirements for the treatment of hyperhomocysteinemia in humans. *J Nutrition* 124(10): 1927–33, 1994.

Aluminum

Alfrey, A.C. Systemic toxicity of aluminum in man. *Neurobiol Aging* 7: 543–4, 1986.

Birchall, J.D., Chappel, J.S. Aluminum, chemical physiology and Alzheimer's disease. *Lancet* 2(8618): 1008–10, 1988.

Campbell, A., Bondy, S. Aluminum induced oxidative events and its relation to inflammation: A role for the metal in Alzheimer's disease. *Cellular and Molecular Biology* (Noisy-le-grand) 46(4): 721–30, June 2000.

Crapper-McLachlan, D., Dalton, A., Kruck, T., et al. Intramuscular desferrioxamine in patients with Alzheimer's disease. *Lancet* 337(8753): 1304–8, August 3, 1991.

Flaten, T. Aluminum as a risk factor in Alzheimer's disease, with an emphasis on drinking water. *Brain Research Bulletin* 55(2): 187–96, May 15, 2001.

Forbes, W., Hill, G. Is exposure to aluminum a risk factor for the development of Alzheimer disease?—Yes. *Archives of Neurology* 55(5): 740–1, May 1998.

Graves, A., Rosner, D., Echeverria, D., et al. Occupational exposures to solvents and aluminum and estimated risk of Alzheimer's disease.

Occupational and Environmental Medicine 55(9): 627–33, September 1998.

Rondeau, V., Commenges, D., Jacqmin-Gadda, H., et al. Relation between aluminum concentrations in drinking water and Alzheimer's disease: An 8-year follow-up study. *American Journal of Epidemiology* 152(1): 59–66, July 1, 2000.

van Rensberg, S.J., Daniels, W.M., Potocnik, F.C., et al. A new model for the pathophysiology of Alzheimer's disease: Aluminum toxicity is exacerbated by hydrogen peroxide and attenuated by an amyloid protein fragment and melatonin. *S Afr J Med* 87(9): 1111–5, 1997.

Weiner, M.A. Evidence points to aluminum's link with Alzheimer's disease. *Townsend Letter for Doctors* 124: 1103, 1993.

Hyperbaric Oxygen Therapy

Anonymous. The Canadian Study of Health and Aging: Risk factors for Alzheimer's disease in Canada. *Neurology* 44: 2073–80, 1994.

Boers, G.H.J., Smals, A.G.H., Trijbels, F.J.M., et al. Heterozygosity for homocystinuria in premature peripheral and cerebral occlusive arterial disease. *N Engl J Med* 313: 709–15, 1985.

Greib, P., Ryba, M.S., Sawicki, J., et al. Oral coenzyme Q10 administration prevents the development of ischemic brain lesions in a rabbit model of symptomatic vasospasm. *Acta Neuropathol (Berl)* 4: 363–8, 1997.

Hayakawa, M. Comparative efficacy of Vinpocetine, pentoxifylline and nicergoline on red blood cell deformability. *Arzneimittelforschung* 42: 108–10, 1992.

Jain, K.K. *Textbook of Hyperbaric Medicine.* 2nd ed. Seattle: Hogrefe and Huber, 1996.

Kang, S.S., Wong, P.W., Malinow, M.R. Hyperhomocysteinemia as a risk factor for occlusive vascular disease. *Ann Rev Nutr* 12: 279–98, 1992.

Kukull, W.A., et al. Solvent exposure as a risk factor for Alzheimer's disease: A case control study. *Am J Epidemiol* 141: 1059–71, 1995.

McCully, K.S. Vascular pathology of homocysteinemia: Implications for the pathogenesis of arteriosclerosis. *Am J Pathol* 56: 111–28, 1969.

Neubauer, R., End, E. Hyperbaric oxygenation as an adjunct therapy in strokes due to thrombosis. *Stroke* 11: 297, 1992.

Neubauer, R., Walker, W. In *Hyperbaric Oxygen Therapy.* Garden City, N.Y.: Avery, 1998.

Olah, V.A., Balla, G., Balla, J., et al. An in vitro study of the hydroxyl scavenger effect of caviton. *Acta Paediatr Hung* 30: 309–16, 1990.

Osawa, M., Maruyama, S. Effects of TCV-3B (Vinpocetine) on blood viscosity in ischemic cerebrovascular disease. *Ther Hung* 33: 7–12, 1985.

Otomo, E., Atarashi, J., Araki, G., et al. Comparison of Vinpocetine with ifenprodil tartrate and dihydroergotoxine mesylate treatment and results of long-term treatment with Vinpocetine. *Curr Therapeut Res* 37: 811–21, 1985.

Robinson, R. Is adaptive plasticity the brain's normal response to injury? *Neurology Reviews* 7(7): 1999.

Selhub, J., Jacques, P.F., Bostom, A.G., et al. Association between plasma homocysteine concentrations and extracranial carotid-artery stenosis. *N Engl J Med* 332: 286–91, 1995.

Selhub, J., Jacques, P.F., Wilson, P.W.F., et al. Vitamin status and intake as primary determinants of homocysteinemia in an elderly population. *JAMA* 270: 2693–8, 1993.

Tamaki, N., Kusunoki, T., Matsumoto, S. The effect of Vinpocetine on cerebral blood flow in patients with cerebrovascular disease. *Ther Hung* 33: 13–21, 1985.

Pesticides

Rajput, Ali H. Environmental toxins accelerate Parkinson's disease onset. *Neurology* 56: 4–5, 2001.

Stephenson, J. Exposure to home pesticides linked to Parkinson disease. *JAMA* 283 (23): 3055–6, June 21, 2000.

Alcohol

Ruitenberg, A., van Swieten, J.C., Witteman, J.C.M., et al. Alcohol consumption and the risk of dementia: The Rotterdam Study. *Lancet* 359: 281–6, 2002.

Truelson, T., Thudium, D., Gronbaek, M. Amount and type of alcohol and risk of dementia. *Neurology* 59: 1313–9, 2002.

Vinpocetine

Gaal, L., Molnar, P. Effect of vinpocetine on noradrenergic neurons in rat locus coeruleus. *Eur J Pharmacol* 187(3): 537–9, 1990.

Hayakawa, M. Comparative efficacy of vinpocetine, pentoxifylline and nicergoline on red blood cell deformability. *Arzneimittelforschung* 42: 108–10, 1992.

Kidd, P.M. A review of nutrients and botanicals in the integrative management of cognitive dysfunction. *Altern Med Rev* 4(3) June 6, 1999.

Olah, V.A., Balla, G., Balla, J., et al. An in vitro study of the hydroxyl scavenger effect of caviton. *Acta Paediatr Hung* 30: 309–16, 1990.

Osawa, M., Maruyama, S. Effects of TCV-3B (vinpocetine) on blood viscosity in ischemic cerebrovascular disease. *Ther Hung* 33: 7–12, 1985.

Otomo, E., Atarashi, J., Araki, G., et al. Comparison of vinpocetine with ifenprodil tartrate and dihydroergotoxine mesylate treatment and results of long-term treatment with vinpocetine. *Curr Therapeut Res* 37: 811–21, 1985.

Tamaki, N., Kusunoki, T., Matsumoto, S. The effect of vinpocetine on cerebral blood flow in patients with cerebrovascular disease. *Ther Hung* 33: 13–21, 1985.

Multiple Sclerosis

Barnes, M.P., Bates, D., Cartidge, N.E. Hyperbaric-oxygen and multiple sclerosis: final results of a placebo-controlled, double-blind trial. *J Neurol Neurosurg Psychiatry* 50(11): 1402–6, 1987.

Davidson, D.L.W. *Hyperbaric Oxygen Therapy in the Treatment of Multiple Sclerosis.* Report from Action for Research into Multiple Sclerosis, London, England, 1989.

Dworkin, R.H., Bates, D., Millar, J.H.D., et al. Linoleic acid and multiple sclerosis: A reanalysis of three double-blind trials. *Neurology* 34: 1441–5, 1984.

Esparza, M.L., Sasaki, S., Kesteloot, H. Nutrition, latitude and multiple sclerosis mortality: An ecologic study. *Am J Epidemiol* 142(7): 733–77, 1995.

Fischer, B.H., Marks, M., Reich, T. Hyperbaric-oxygen treatment of multiple sclerosis: A randomized, placebo-controlled, double-blind study. *N Engl J Med* 308(4): 181–6, 1983.

Geissler, A., Andus, T., Roth, M., et al. Focal white-matter lesions of the brain in patients with inflammatory bowel disease. *Lancet* 345: 897–8, 1995.

Lauer, K. Diet and multiple sclerosis. *Neurology* 49(suppl 2): S55–61, 1997.

Nieves, J., Cosman, F., Shen, H.J. et al. High prevalence of vitamin D deficiency and reduced bone mass in multiple sclerosis. *Neurology* 44(9): 1687–92, 1994.

Perlmutter, D. Fatigue in multiple sclerosis. *Townsend Letter for Doctors* 148: 48–50, 1995.

Reynolds, E.H. Multiple sclerosis and vitamin B12 metabolism. *J Neuroimmun* 40: 225–30, 1992.

Rudnick, R.A., Cohen, J.A., Weinstock-Guttman, B., et al. Management of multiple sclerosis. *N Engl J Med* 337(22): 1604–67, 1997.

Rudnick, R.A., Ransohoff, R.M., Herndon, R.M. Multiple sclerosis and other myelin disorders. In *Clinical Neurology*, vol. 3, edited by R.J. Joint Rochester, N.Y.: Lippincott Williams and Wilkins, 1998.

Swank, R.L. Multiple sclerosis: The lipid relationship. *A J Clin Nutr* 48: 1387–93, 1988.

Swank, R.L., Lerstad, O., Strom, A. Multiple sclerosis in rural Norway: Its geographic and occupational incidence in relation to nutrition. *N Eng J Med* 246: 721–8, 1952.

Lifestyle and Diet

Abbot, R.D., Ross, G.W., White, L.R., et al. Midlife adiposity and the future risk of Parkinson's disease. *Neurology* 59: 1051–7, 2002.

Golbe, L.I., Farrell, T.M., David, P.H. Case-control study of early life dietary factors in Parkinson's disease. *Arch Neurol* 45(12): 1350–3, 1988.

Hellenbrand, W., Seidler, A., Boeing, H., et al. Diet and Parkinson's disease I: A possible role for the past intake of specific foods and food groups. *Neurology* 47: 636–43, 1996.

Luchsinger, J.A., Tang, Ming-Xing, Shea, S., et al. Caloric intake and the risk of Alzheimer disease. *Arch Neurol* 59: 1258–63, 2002.

Mortensen, E.L., Michaelsen, K.F., Sanders, S.A. The association between duration of breast feeding and adult intelligence. *JAMA* 287(18): 2365–71, 2002.

Rosenberg, I.H., Miller, J.W. Nutritional factors in physical and cognitive functions of elderly people. *Am J Clin Nutr* 55: 1237S–43S, 1992.

Toole, J.F., Jack, C.R. Food (and vitamins) for thought. *Neurology* 58: 1449–50, 2002.

Ginkgo

Curtis-Prior, P., et al. Therapeutic value of ginkgo biloba in reducing symptoms of decline in mental function. *Journal of Pharmacy and Pharmacology* 51: 535–41, 1999.

Hindmarch, I. Activity of ginkgo biloba extract on short-term memory. *Presse Medicale* 15(31): 1592–4, 1986.

Itil, T.M., et al. Central nervous system effects of ginkgo biloba, a plant extract. *American Journal of Therapeutics* 3: 63–73, 1996.

Le Bars, P., Katz, M.M., Berman, N., et al. A placebo-controlled, double-blind randomized trial of an extract of ginkgo biloba for dementia. *JAMA* 278(16): 1327–32, 1997.

Pidoux, B. Effects of ginkgo biloba on cerebral functional activity: Results of clinical and experimental studies. *Presse Medicale* 15(31): 1588–91, 1986.

Essential Fatty Acids–DHA

Arvindakshan, M., et al. Supplementation with a combination of omega-3 fatty acids and antioxidants (vitamins E and C) improves the outcome of schizophrenia. *Schizophr Res* 62(3): 195–204.

Conner, W.E., et al. Dietary effects on brain fatty acid composition: the reversibility of n-3 fatty acid deficiency and turnover of docosahexaenoic acid in the brain, erythrocytes, and plasma of rhesus monkeys. *J Lipid Res* 31(2): 237–47, 1990.

de Wilde, M.C., et al. The effect of n-3 polyunsaturated fatty acid–rich diets on cognitive and cerebrovascular parameters in chronic cerebral hypoperfusion. *Brain Res* 947(2): 166–73, August 15, 2002.

Favrelière, S., et al. DHA-enriched phospholipid diets modulate age-related alterations in rat hippocampus. *Neurobiol Aging* 24(2): 233–43, December 25, 2002.

Gamoh, S., Hashimoto, M., et al. Chronic administration of docosahexaenoic acid improves reference memory-related learning ability in young rats. *Neuroscience* 93(1): 237–41, 1999.

Hoffman, D.R., et al. Visual function in breast-fed term infants weaned to formula with or without long-chain polyunsaturates at 4 to 6 months: A randomized clinical trial. *J Pediatr* 142(6): 669–77, July 3, 2003.

Itokazu, N., et al. Bidirectional actions of docosahexaenoic acid on hippocampal neurotransmissions in vivo. *Brain Res* 862(1–2): 211–6, May 9, 2000.

Morris, M.C., et al. Consumption of fish and n-3 fatty acids and risk of incident Alzheimer disease. *Arch Neurol* 60: 940–6, 2003.

Newman, P.E. Could diet be one of the causal factors of Alzheimer's disease? *Med Hypotheses* 39(2): 123–6, October 1992.

Pischon, T., et al. Habitual dietary intake of n-3 and n-6 fatty acids in relation to inflammatory markers among US men and women. *Circulation* 108(2): 155–60, June 25, 2003.

Simopoulos, A.P., et al. Omega-3 fatty acids in inflammation and autoimmune diseases. *J Am Coll Nutr* 21(6): 495–505, December 14, 2002.

Tanskanen, A., et al. Fish consumption, depression, and suicidality in a general population. *Arch Gen Psychiatry* 58(5): 512–3, May 16, 2001.

ApoE4

Chapman, J., et al. The effects of APOE genotype on age at onset and progression of neurodegenerative diseases. *Neurology* 57: 1482–5, 2001.

Cohen, R.M., Small, C., Lalonde, F. Effect of apolipoprotein E genotype on hippocampal volume loss in aging healthy women. *Neurology* 57: 2223–8, 2001.

Corder, E.H., et al. Gene dose of apolipoprotein E type 4 allele and the risk of Alzheimer's disease in late onset families. *Science* 261: 921–3, 1993.

Dik, M.G., Jonker, C., Comijs, H.C. Memory complaints and APOE-e4 accelerate cognitive decline in cognitively normal elderly. *Neurology* 57: 2217–22, 2001.

Farlow, M.R. Alzheimer's disease: Clinical implications of the apolipoprotein E genotype. *Neurology* 48(5): S30–4, 1997.

Masliah, E. Alterations in apolipoprotein E expression during aging and neurodegeneration. *Prog Neurobiol* 50(5–6): 493–503, 1996.

Saunders, A.M., et al. Association of apolipoprotein E allele E4 with late-onset familial and sporadic Alzheimer's disease. *Neurology* 43: 1467–72, 1993.

Strittmatter, W.E., et al. Apolipoprotein E: High avidity binding to beta-amyloid and increased frequency of type 4 allele in late-onset familial Alzheimer disease. *Proc Natl Acad Sci* 90: 1977–81, 1993.

Coenzyme Q10

Backes, J.M., Howard, P.A. Association of HMG-CoA reductase inhibitors with neuropathy. *Ann Pharmacother* 1–29 37(2) 274–78, 2003.

Beal, M.F. Mitochondria, oxidative damage, and inflammation in Parkinson's disease. *Ann N Y Acad Sci* 991: 120–31, July 2003.

Matthews, R.T., et al. Coenzyme Q10 administration increases brain mitochondrial concentrations and exerts neuroprotective effects. *Proc Natl Acad Sci USA* 95(15): 8892–7, July 21, 1998.

Mortensen, S.A., Leth, A., Agner, E. Dose-related decrease of serum coenzyme Q10 during treatment with HMG-CoA reductase inhibitors. *Mol Aspects of Med* 18(suppl): S137–44, 1997.

Sandhu, J.K., et al. Molecular mechanisms of glutamate neurotoxicity in mixed cultures of NT2-derived neurons and astrocytes: Protective effects of coenzyme Q10. *J Neurosci Res* 72(6): 691–703, May 30, 2003.

Shults, C.W., Beal, M.F., Fontaine, K. et al. Absorption, tolerability and effects on mitochondrial activity of oral coenzyme Q10 in Parkinsonian patients. *Neurology* 50: 793–5, 1998.

Shults, C.W., Haas, R.H., Passov, D., Beal, M.F. Coenzyme Q10 levels correlate with the activities of complexes I and II/III in mitochondria from Parkinsonian and non-Parkinsonian subjects. *Ann Neurol* 42: 261–4, 1997.

Shults, C.W., Oakes, D., Kieburtz, K., et al. Effects of coenzyme Q10 in early Parkinson disease. *Arch Neurol* 59: 1541–50, 2002.

Phosphatidylserine

Amenta, F., et al. Treatment of cognitive dysfunction associated with Alzheimer's disease with cholinergic precursors. Ineffective treatments or inappropriate approaches? *Mech Ageing Dev* 122(16): 2025–40, November 2001.

Benton, D., et al. The influence of phosphatidylserine supplementation on mood and heart rate when faced with an acute stressor. *Nutr Neurosci* 4(3): 169–78, 2001.

Cenacchi, T., et al. Cognitive decline in the elderly: A double-blind, placebo-controlled multicenter study on efficacy of phosphatidylserine administration. *Aging* (Milano) 5(2): 123–33, April 1993.

Crook, T.H., Tinklenberg, J., Yesavage, J. Effects of phosphatidylserine in age-associated memory impairment. *Neurology* 41: 644–9, 1991.

Crook, T.H., et al. Effects of phosphatidylserine in Alzheimer's disease. *Psychopharmacol Bull* 28(1): 61–6, 1992.

Delwaide, P.J., et al. Double-blind randomized controlled study of phosphatidylserine in senile demented patients. *Acta Neurol Scand* 73(2): 136–40, February 1986.

Engel, R.R. Double-blind crossover study of phosphatidylserine vs. placebo in patients with early dementia of the Alzheimer type. *Eur Neuropsychopharmacol* 2(2): 149–55, June 1992.

Funfgeld, E.W., et al. Double-blind study with phosphatidylserine (PS) in Parkinsonian patients with senile dementia of Alzheimer's type (SDAT). *Prog Clin Biol Res* 317: 1235–46, 1989.

Heiss, W.D., et al. Activation PET as an instrument to determine therapeutic efficacy in Alzheimer's disease. *Ann NY Acad Sci* 24(695): 327–31, September 1993.

Kidd, P.M. A review of nutrients and botanicals in the integrative management of cognitive dysfunction. *Altern Med* 4(3): 144–61, June 1999.

Maggioni, M., et al. Effects of phosphatidylserine therapy in geriatric patients with depressive disorders. *Acta Psychiatr Scand* 81: 265–70, 1990.

Monteleone, P., et al. Blunting by chronic phosphatidylserine administration of the stress-induced activation of the hypothalamo-pituitary-adrenal axis in healthy men. *Eur J Clin Pharmacol* 42(4): 385–8, 1992.

Monteleone, P., et al. Effects of phosphatidylserine on the neuroendocrine response to physical stress in humans. *Neuroendocrinology* 52(3):243–8, September 1990.

Palmieri, G., et al. Double-blind controlled trial of phosphatidylserine in patients with senile mental deterioration. *Clin Trials J* 24: 73–83, 1987.

Suzuki, S., et al. Oral administration of soybean lecithin transphosphatidylated phosphatidylserine improves memory impairment in aged rats. *J Nutr* 131(11): 2951–6, November 2001.

Lipoic Acid

Bast, A., et al. Lipoic acid: A multifunctional antioxidant. *Biofactors* 17(1–4): 207–13, August 5, 2003.

Chen, H.J., et al. Biological and dietary antioxidants protect against DNA nitration induced by reaction of hypochlorous acid with nitrite. *Arch Biochem Biophys* 415(1): 109–16, July 1, 2003.

Koenig, M.L., et al. In vitro neuroprotection against oxidative stress by pre-treatment with a combination of dihydrolipoic acid and phenyl-butyl nitrones. *Neurotox Res* 5(4): 265–72, July 2, 2003.

Lovell, M.A., et al. Protection against amyloid beta peptide and iron/hydrogen peroxide toxicity by alpha lipoic acid. *J Alzheimers Dis* 5(3): 229–39, August 5, 2003.

Marangon, K., Devaraj, S., Tirosh, O., et al. Comparison of the effect of a-lipoic acid and a-tocopherol supplementation on measures of oxidative stress. *Free Radic Biol Med* 27(9/10): 1114–21, 1999.

Marriage, B., et al. Nutritional cofactor treatment in mitochondrial disorders. *J Am Diet Assoc* 103(8): 1029–38, August 2, 2003.

Packer, L., et al. Neuroprotection by the metabolic antioxidant lipoic acid. *Free Radic Biol Med* 22 (1/2): 359–78, 1997.

Tirosh, O., et al. Neuroprotective effects of lipoic acid and its positively charged amide analogue. *Free Radic Biol Med* 26 (11/12): 1418–26, 1999.

N-Acetyl-Cysteine

Chen, H.J., et al. Biological and dietary antioxidants protect against DNA nitration induced by reaction of hypochlorous acid with nitrite. *Arch Biochem Biophys* 415(1): 109–16, July 1, 2003.

De Quay, B., et al. Glutathione depletion in HIV-infected patients: Role of cysteine deficiency and effect of oral N-acetylcysteine. *AIDS* 6: 815–9, 1992.

Medina, S., et al. Antioxidants inhibit the human cortical neuron apoptosis induced by hydrogen peroxide, tumor necrosis factor alpha, dopamine and beta-amyloid peptide 1–42. *Free Radic Res* 36(11): 1179–84, November 2002.

Neal, R., et al. Antioxidant role of N-acetyl cysteine isomers following high dose irradiation. *Free Radic Biol Med* 34(6): 689–95, March 15, 2003.

Pahan, K., Sheikh, G.S., Nmboodiri, A.M.S., et al. N-acetyl cysteine inhibits induction of NO production by endotoxin or cytokine stimulated rat peritoneal macrophages, C6 glial cells and astrocytes. *Free Radic Biol Med* 24(1): 39–48, 1998.

Poliandri, A.H., et al. Cadmium induces apoptosis in anterior pituitary cells that can be reversed by treatment with antioxidants. *Toxicol Appl Pharmacol* 190(1): 17–24, July 1, 2003.

Willis, M.D., et al. N-Acetylcysteine in the treatment of human arsenic poisoning. *J Am Board Fam Pract* 3: 293–6, 1990.

Index

ABC drugs (Avonex, Betaseron, Copaxone), 230–31
Abuse, in childhood, 35
Acetaminophen (Panadol, Tylenol, Aspirin-Free Anacin), 12, 24, 27, 41, 56
  and glutathione, 54–55, 117–18, 210
Acetylcholine, 22, 101–2, 133, 206, 211
Acetyl-L-carnitine, 101–3, 178
  recommended uses, 125, 126, 200, 203, 214, 225, 237, 244, 248
    Acetylsalicylic acid. *See* Aspirin
Activela (estrogen), 15, 50
Adenosine triphosphate (ATP), 43
Advanced glycation end products (AGES), 38
Advil (ibuprofen), 15, 50
Aerobid (flunisolide), 14, 49
Aerolate (theophylline), 14, 49
Age, Brain Audit, 9
AGES (advanced glycation end products), 38
Aging, and brain, 4, 25
Alcohol, 27, 72, 84, 112
  and acetaminophen, 56
  and sleep disorder, 131
  and stroke, 194–95
Aldactazide (hydroclorothiazide), 13, 45, 49
Aleve (naproxen), 15, 50
Ali, Muhammad, 216

Alpha lipoic acid, 103–4, 178
  recommended uses, 125, 126, 200, 203, 213, 223, 225, 237, 241, 244, 248
ALS (amyotrophic lateral sclerosis), 143, 239–45
Aludrox (Aluminum hydroxide), 12, 57
Aluminum, 28, 56–59, 149–51, 210
Aluminum hydroxide (Gaviscon, Aludrox, Di-Gel, Gelusil, Maalox, Magalox, Mylanta), 12, 57
Aluminum lauryl sulfate, 150
Alzheimer's disease, 23, 28, 38–39, 204–14
  acetaminophen and, 55
  aluminum and, 56–57, 150
  anti-inflammatory drugs and, 48
  B vitamins and, 188
  carnitine and, 102–3
  Co-Q10 and, 109
  CRP and, 184
  DHA and, 64
  electromagnetic radiation and, 155
  exercise and, 30, 136
  genetics and, 36–37, 181–82
  glutathione and, 118
  homocysteine and, 115, 187
  lipid peroxide and, 171
  melatonin and, 132

Carbamazepine (Tegretol), 14, 49
Carbidopa (Sinemet), 14, 47, 50
Carnitine, 101–3
Carotid artery, blockage of, 195
Catapres (clonidine), 13, 45
Cauliflower, 220
CDSA (comprehensive digestive stool
    analysis), 230
Celebrex (celecoxib), 15, 47, 50, 241, 242,
    243
Celiac disease, 234–36
Cell membranes, 20, 22
    acetylcholine and, 206
    DHA and, 111
    phosphatidylserine (PS) and, 119
    trans-fatty acids and, 65, 186
    vitamin E and, 113
Cell phones, 29–30, 141, 154, 155–57, 291
Celusil, 12
Ceramic dishes, lead glazed, 152
Cerebyx (fosphenytoin), 14, 49
Chelating agents, 104, 118
Chelation therapy, 33, 149, 152, 291
Chemical toxins, 142
Chemotherapy, antioxidants and, 124
Chicago Health and Aging Project, 113
Chicken, white meat, 212
    See also Recipes
Childhood abuse, and Alzheimer's disease,
    35
Children, lead and, 151
"Chinese restaurant syndrome," 153
Chlamydia pneumoniae, 229, 237
Chlorine, 72
Cholesterol, eggs and, 73
Cholesterol-lowering drugs, 4, 13, 41, 43
    and B vitamins, 50
    and Co-Q10, 46
    and free radicals, 24
Cholestyramine (Colestid), 13, 50
Chronic stress, 33–34, 133
Cimetidine (Tagamet), 12, 48
Citronella, 145
Clock radios, 33, 157
Clonidine (Catapres), 13, 45
Cobalamin (vitamin B12), 51–54, 104,
    106–7, 179, 180, 187
    and homocysteine, 178
    injections, 237
    recommended uses, 125, 126, 200, 203,
        214, 225, 238, 244, 247, 248, 249
Cocaine, and brain function, 31

Cockroaches, control of, 145–46
Codeine, 12, 48
Coenzyme-Q10. *See* Co-Q10
Coffee, 72, 84
Cognex (tacrine), 211–12
Cognitive function, decreased, DHA and, 64
Cold remedies, over-the-counter, 12
Colestid (cholestyramine), 13, 50
Combinpatch (estrogen), 15, 50
Combipres (hydroclorothiazide), 13, 45, 49
Companion plantings, 144–45
Complementary medicine, 215
Comprehensive digestive stool analysis
    (CDSA), 230
Computers, 33, 141, 154, 157
Computer software, mental enhancement,
    291
Concentration, difficulties of, 19–20
Condiments, 84–85, 87
Confusion, B vitamins and, 106
Constipation, in Parkinson's disease, 224–25
Consumer Product Safety Commission
    (CPSC), 139–40
Cooking oils, 63
Cookware, aluminum, 150, 151
Copaxone, 230–31
Copper pipes, 152
Co-Q10 (Coenzyme-Q10), 24, 28, 41,
    108–11, 178, 197, 226, 241, 287
    medications and, 42–46
    recommended uses, 124, 125, 126, 200,
        203, 213, 225, 237, 244, 247, 248
Corgard (nadolol), 13, 45
Corn oil, 64
Coronary artery disease (CAD), 39, 195
Corticosteroid drugs, 14, 51, 133
Cortisol, 102
Costs
    of Alzheimer's disease, 205
    of stroke, 193
Coumadin, vinpocetine and, 120, 121, 124,
        125, 126, 179, 200, 203, 214, 225,
        238, 245, 248, 249
Counseling, for stress, 135
COX-2 enzyme, 241
C-reactive protein (CRP), 183–84, 210
Creatine monohydrate, 244
Crook, Thomas, 118–19
CRP (C-reactive protein), 183–84, 210
Cruciferous vegetables, 70, 220
C vitamin, 107–8, 178, 199
    and antidiabetic drugs, 124

Neurology journal, statin study, 44
Neurons, 20, 219
  idling, 197, 198
Neurotoxins, 141–58
  lead, 151–52
  mercury, 146–48
Neurotransmitters, 22–23, 43, 102, 178
  corticosteroids and, 133
  depression and, 38
  excitotoxins and, 153
  sleep and, 129–30
New England Journal of Medicine:
  ALS study, 233
  Alzheimer's disease studies, 113, 209
Nexium, and B vitamins, 47
Niacin (vitamin B3), 51–53, 104, 179, 187
  recommended uses, 125, 126, 200, 203,
    214, 225, 237, 244, 247, 248, 249
Nicotine, 130, 131, 194–95
Nizatidine (Axid), 12, 49
Nonaspirin pain relievers, 4, 12, 55, 56
Nonsteroidal antiflammatory drugs. *See*
    NSAIDS
Nordic Natural products, 286–87
Norpramin (desipramine), 12, 45
Nortriptyline (Aventil, Pamelor), 12, 45
Novantrone, 230–31
NSAIDs (nonsteroidal antiflammatory
    drugs), 15, 27, 47–48
  and Alzheimer's disease, 210
  and B vitamins, 50
Number recall, exercise for, 161–63
Nut butters, 79
Nutrient deficiency, 235
Nutrient depletion, drug-induced, 41
Nutrition, 61–99
  stroke therapy, 199–200
Nutritional supplements. *See* Supplements
Nuts, 63, 64, 79, 212

Obesity, 24, 37–38, 76, 199
  and stroke, 194–95
Occupations, and risk of ALS, 240
Ohio State University, stress study, 134
Oils, 80
Olive oil, 63, 212
Omega 3 fatty acids, 64–68, 74, 112
Omega 6 fatty acids, 64, 77
Omeprazole (Prilosec), 12
Oral contraceptives, 15, 50, 195
Oral inhalers, 14, 49
Orasone (prednisone), 14, 51

Organic Gardening, 145
Organic gardening information, 290
Organic produce, 32, 62, 67, 70–71, 146
  fruits, 82
  vegetables, 81
Ortho-Novum (estrogen), 15, 50
Ortho Tri-Cyclen (estrogen), 15, 50
Osteoporosis, 235
  estrogen substitutes for, 15, 50
Ovcon (estrogen), 15, 50
Over-the-counter drugs. *See* Medications
Ovral (estrogen), 15, 50
Oxidation, 22
Oxidative Stress Test, 172
Oxygen, stroke therapy, 197–98
OxyStress Test, 174–75

Pain relievers, 12, 41, 48
  nonaspirin, 4, 55, 56
Paint, lead-based, 152
Pamelor (nortriptyline), 12, 45
Panadol (acetaminophen), 12, 56
Pans, aluminum, 150, 151
Parent, loss of, and brain function, 35
Parkinson, James, "An Essay on the Shaking
    Palsy," 217
Parkinson's disease, 23, 28, 43, 215–21
  cocaine and, 31
  Co-Q10 and, 109
  diagnosis, 217–19
  glutathione and, 55, 222–24
  homocysteine and, 179
  medications for, 47
  nutrition advice, 224–26
  risk of, 32, 37, 39
    head injury and, 139
    pesticides and, 143
  surgery for, 221–22
Peanut oil, 64
Pedometers, 137
Pelton, Ross, *Drug-Induced Nutrient
    Depletion Handbook,* 42
Pepcid (famotidine), 12, 48
Percodan, 12, 48
Peripheral neuropathy, 44
Perlmutter, David, products designed by,
    287–88
Perlmutter Health Center, 5, 109, 211
  and ALS, 240, 242
  and multiple sclerosis, 230, 231, 233
  and Parkinson's disease, 215–16, 222–24, 226
  stroke recovery protocol, 197

Thiamine (vitamin B1), 51–53, 104, 105, 179, 187
  recommended uses, 125, 126, 200, 203, 214, 225, 237, 244, 247, 248, 249
TIA (transient ischemic attack), 196
Tilapia, 68, 147
  See also Recipes
TMG (trimethylglycine), 52, 53, 54, 180
Tobacco, and brain function, 30–31
Tofranil (imipramine), 12, 45
Tofu, 67, 79
  See also Recipes
Tolazemide (Tolinase), 13, 46
Tooth fillings, mercury in, 118, 148–49
Toprol (metoprolol), 13, 46
Torsemide (Demadex), 13, 49
Toxins, environmental, 5, 31–33, 141–58
  and ALS, 240
  and Alzheimer's disease, 210
  in foods, 70–72
  glutathione and, 54
  and Parkinson's disease, 219
Trans-fatty acids, 26, 64, 65–66, 87, 112, 186, 212
Transient ischemic attack (TIA), 196
Tremor, of Parkinson's disease, 218
Triamcinolone (Azmacort), 14, 49
Triamterene (Dyrenium, Maxide, Dyazide), 13, 49
Trichogramma wasps, 144
Trimethoprim (Bactrim, Septra), 14, 48
Trimethylglycine (TMG), 52, 53, 54, 180
Tufts University, homocysteine study, 195
Tums, 58
Tuna, 147
Turkey, 212
Tylenol (acetaminophen), 12, 24, 56
Tyrosine, 108

Ultraviolet rays, 154
Underarm deodorants, 5
United States Department of Agriculture (USDA), and organic produce, 70, 71
University of Cincinnati Medical Center, and aluminum, Alzheimer's study, 150
University of Kansas, ALS research, 243
University of Michigan, cocaine study, 31
University of Southern California School of Medicine, occupations and Alzheimer's study, 155

University of Toronto, Alzheimer's study, 57
Urine, detoxification function, 142

Valproic acid (Depakote, Depakene), 14, 50
Vanceril (beclomethasone), 14, 49
Vanderbilt School of Medicine, *chlamydia* study, 229
Vascular dementia, 47, 107–8, 201–3
  vinpocetine and, 119–20
Vaseretic (hydroclorothiazide), 13, 45, 49
Vegetable oils, 64
Vegetables, 27, 68, 69–70, 77, 81–82
  cruciferous, 220
  green leafy, 64, 178, 199, 212
  See also Recipes
Vegetarian diet, and B vitamins, 105, 106
Vespro Free Radical Test, 174–75
Vinca minor, 120
Vinpocetine, 119–21, 178, 199
  recommended uses, 124, 125, 126, 200, 202, 203, 214, 225, 237, 244, 245, 248, 249
Visken (pindolol), 13, 46
Vitamins. *See* Supplements; letter name of vitamin
Vivactil (protriptyline), 12, 45
Vons frozen waffles, 68

Walking, 137
Walnut oil, 74, 212
Walnuts, 64, 68
  See also Recipes
Water, 5, 32, 72, 84, 150, 152, 290, 291
Weight loss, gluten intolerance and, 235
Whole grains, 83
Wine, 72, 84
Women, mercury levels in, 147–48

Yeast overgrowth, 229–30
Yoga, 136–37, 288

Zabeta (bisoprolol), 13
Zantac, Zantac 75 (ranitidine), 12, 48
Zarontin (ethosuximide), 14, 49
Zeaxanthin, 70
Zebeta (bisoprolol), 45
Zestoretic (hydroclorothiazide), 13, 45, 49
Zocor (simvastatin), 13, 44, 46

About the Authors

David Perlmutter, M.D., FACN, a board certified neurologist and founder of the Perlmutter Health Center, is a leader in the field of complementary medicine. His scientific writing has appeared in *The Journal of the American Medical Association*, the *Journal of Neurosurgery*, and *The Southern Medical Journal*. He Lives in Naples, Florida.

Carol Colman is the coauthor of numerous bestselling health books, She lives in Larchmont, New York,

For more information, please visit www. BetterBrainBook.com